How To Be
An Intellectually Fulfilled Atheist

(Or Not)

How To Be
An Intellectually Fulfilled Atheist

(Or Not)

William A. Dembski

Jonathan Wells

Wilmington, Delaware

Dembski, William A., 1960–

    How to be an intellectually fulfilled atheist (or not) / William A. Dembski, Jonathan Wells.—1st ed.—Wilmington, Del.: ISI Books, c2008.

        p. ; cm.
        ISBN: 978-1-933859-84-2
        Edited and reorganized from the last chapter of "The design of life : discovering signs of intelligence in biological systems" by William A. Dembski and Jonathan Wells. (Dallas : Foundation for Thought and Ethics, c2008.)
        Includes bibliographical references.

        1. Intelligent design (Teleology) 2. Atheism. 3. Evolution (Biology) Religious aspects—Christianity. 4. Religion and science I. Wells, Jonathan, Ph. D. II. Dembski, William A., 1960–. Design of life. III. Title.

BL263 .D464 2008   2008934995
213—dc22   0810

Book design by Meghan Duke

ISI Books
Intercollegiate Studies Institute
P.O. Box 4431
Wilmington, Delaware 19807-0431
www.isibooks.org

Manufactured in the United States of America

Contents

"There is superstition in science quite as much as there is superstition in theology, and it is all the more dangerous because those suffering from it are profoundly convinced that they are freeing themselves from all superstition. No grotesque repulsiveness of medieval superstition, even as it survived into nineteenth-century Spain and Naples, could be much more intolerant, much more destructive of all that is fine in morality, in the spiritual sense, and indeed in civilization itself, than that hard dogmatic materialism of to-day which often not merely calls itself scientific but arrogates to itself the sole right to use the term. If these pretensions affected only scientific men themselves, it would be a matter of small moment, but unfortunately they tend gradually to affect the whole people, and to establish a very dangerous standard of private and public conduct in the public mind."

—Theodore Roosevelt
"The Search for Truth in a Reverent Spirit"
Outlook, December 2, 1911

Introduction: Atheism's Quest for Intellectual Fulfillment

Although atheism might have been logically tenable before Darwin," writes Richard Dawkins, "Darwin made it possible to be an intellectually fulfilled atheist."[1] Certainly, many people look to Darwin's theory of evolution to underwrite atheism. Cornell's Will Provine has even remarked that Darwinian evolution "is the greatest engine of atheism ever invented."[2] But the question is not simply whether Darwin's theory underwrites atheism. Foolish arguments that can be easily dismissed do that. The question is whether, based on a careful examination of available evidence, Darwin's theory underwrites an intellectually fulfilled form of atheism. This little book shows that atheism must seek intellectual fulfillment elsewhere.

The central claim of Darwin's theory is that unguided material processes (natural selection acting on random variations) account for the emergence of all biological complexity and diversity *once life already exists.* Darwin's theory presupposes life. But in that case, the theory does not provide a complete self-contained materialist explanation of life, as required for an intellectually fulfilled atheism. Rather, the theory remains open at one end until an adequate materialist explanation of life's origin is found.

Darwin did not fully appreciate the challenge that life's origin posed to his theory. In his day, most scientists regarded the problem of life's origin as simple and straightforward. The basis of life itself—the cell—was regarded as uncomplicated, essentially a blob of jelly enclosed by a membrane. Darwin took the spontaneous generation of first life from nonliving materials for granted. For instance, he speculated whether in some warm little pond, with all sorts of ammonia and phosphoric salts, lights, heat, electricity, etc. present, that a proteine [*sic*] compound was chemically formed ready to undergo still more complex changes. . . . [3]

That's why he wrote *On the Origin of Species* rather than *On the Origin of Life*. The challenge, he thought, was to explain the diversification of life into species, not its origin.

Nonliving matter was believed to have the power to organize itself into something as simple as a cell. But was the cell really so simple? Was Darwin right to dismiss this loophole in his theory? Heinrich Georg Bronn, who translated the *Origin of Species* into German in 1860 (just a year after its release in 1859), understood what was at stake:

> [If] organic matter could be generated from inorganic matter and if a faultless proof could be provided to show that organic species can arise in the manner suggested by Darwin, then his theory would receive the strongest possible support in the shortest possible time. . . . Thus, we will no longer be forced to seek recourse in personal acts of creation which fall outside the scope of natural law. Once we are in possession of this advantage, we need no longer doubt as before in the possibility of later discoveries gradually filling in the enormous gaps which now confront us in the series of plant and animal forms and impede our complete consent. However, as long as this is not possible, Darwin's theory is as improbable as ever, since it brings us no closer to the solution of the great problem of creation. Though it is conceivable, it still remains undecided whether all organisms from the simplest filament to those that are intricately constructed like the butterfly, snake, horse, etc. can be the production of just blind forces! [4]

In other words, if blind forces are to explain life, they must explain both the origin of life as well as its subsequent development. Darwin's theory attempts to explain only its subsequent development.

But what if blind forces give no evidence of explaining the origin of life? What if, further, such forces work against a purely material origin of life? Then we must consider nonmaterial explanations of life's origin. Darwin might have tacitly agreed. Throughout his *Origin of Species,* Darwin constantly argued that blind material forces (notably natural selection) could substitute for the action of a creative intelligence. But what if matter is no substitute for mind?

"It was Darwin's greatest accomplishment," writes Francisco Ayala, "to show that the directive organization of living beings can be explained as the result of a natural process, natural selection, without any need to resort to a Creator or other external agent."[5] Yet without a materialist account of life's origin, Darwin accomplished nothing of the sort. Rather, he left out the thing that most needed to be explained, namely, how blind material forces could organize the first cell. An intellectually fulfilled atheist must answer this question satisfactorily.

This book shows that Dawkins is wrong. Life's origin poses insuperable difficulties to unguided material processes, so intellectual fulfillment remains for atheism but an elusive dream. This book is drawn from a larger work titled *The Design of Life: Discovering Signs of Intelligence in Biological Systems* (Foundation for Thought and Ethics, 2007). There, we critically assess contemporary evolutionary biology, showing that blind material forces adequately explain neither the origin of life nor its subsequent history. We even go further, arguing that some form of intention or purpose must figure in any adequate explanation of life. Simply put, we argue that life is intelligently designed.

The weakest link in any materialist understanding of life centers on life's origin. Darwin's theory gives at least the semblance of a coherent materialist explanation for how life may have diversified into many species once it actually existed (though in our view it fails even in this regard). But the origin of life lacks even the semblance of such an explanation. Chapter 8, the last chapter in *The Design of Life,* examines the origin of life. For us, this chapter was a slam-dunk: even if evolutionary biology could refute everything we had done in previous chapters, chapter 8 decisively demonstrates the need for intelligence in explaining life's origin.

Edited and reorganized, and without the pretty pictures, the last chapter of *The Design of Life* has become the present book. We offer it as a refutation of atheism, as a defense of intelligent design, and as a stimulus to read the full case for intelligent design presented in *The Design of Life*.

1

The Problem of Life's Origin

The origin of life is as difficult a problem as exists in science. But what exactly is the problem? And what would it mean to solve it? The mathematician George Polya once quipped, "If you can't solve a problem, then there is an easier problem you can solve: find it."[1] This can be good advice if the easier problem illuminates the original problem. But if the easier problem lulls us into thinking that we have solved, or are on the verge of solving, the original problem when in fact we're clueless, then Polya's advice is bad. And in the case of life's origin, Polya's advice is terrible.

Most of origin-of-life research conveniently redefines the problem of life's origin to make it so easy that origin-of-life researchers think they have a reasonable shot at solving it. In fact, they're kidding themselves. Most origin-of-life research is as relevant to the real problem of life's origin as rubber-band-powered propeller model planes are to the military's most sophisticated stealth aircraft.

Why is this so? Mainly because real life is so much more sophisticated than any of the supposed "forerunners" to life put forward in conventional origin-of-life research that there is no reason to think that this research provides any glimmer of insight into life's actual origin. What, then, is the real problem of life's origin? It is to explain the origin of cells as we currently find them on planet Earth, in their full jaw-dropping complexity.[2]

Magnified several hundred times with an ordinary microscope—the sort that was available in Darwin's day—a living cell is a disappointing sight. It looks like a disordered collection of blobs that unseen turbulent forces toss in all directions. To grasp the reality of life as revealed by contemporary molecular biology, we need to magnify the cell a billion times. At that level of magnification, a typical eukaryotic cell (i.e., cell with a nucleus) is more than ten miles in diameter and resembles a giant spaceship large enough to engulf a sizable city. Here we see an object of unparalleled complexity and adaptive design.

On the surface are millions of openings, like the portholes of a ship, opening and closing to allow a continual stream of materials to flow in and out. As we enter one of these openings, we discover a world of supreme technology and bewildering complexity. We see endless highly organized corridors and conduits branching in every direction from the perimeter of the cell, some leading to the central memory bank in the nucleus and others to assembly plants and processing units.

The nucleus itself is a vast chamber a mile in diameter resembling a geodesic dome. Inside we see, all neatly stacked together in ordered arrays, coiled chains of DNA thousands and even millions of miles in length. This DNA serves as a memory bank to build the simplest functional components of the cell, the protein molecules. Yet proteins themselves are astonishingly complex pieces of molecular machinery. An average protein consists of several hundred precisely ordered amino acids arranged in a highly organized three-dimensional structure.

Robot-like machines working in synchrony shuttle a huge range of products and raw materials along the many conduits to and from all the various assembly plants in the outer regions of the cell. Everything is precisely choreographed. Indeed, the level of control implicit in the coordinated movement of so many objects down so many seemingly endless conduits, all in unison, is mind-boggling.

As we watch the strangely purposeful activities of these uncanny molecular machines, we quickly realize that despite all our accumulated knowledge in the natural and engineering sciences, the task of designing even the most basic components of the cell's molecular machinery, the proteins, is completely beyond our present capacity. Yet the life of the cell depends on the integrated activities of many different protein molecules, most of which work in integrated complexes with other proteins.

In touring the cell, we see that nearly every feature of our own advanced technologies has its analogue inside the cell. The cell's nanotechnology has an elegance and sophistication that puts our best human engineering to shame. The conclusion that life is designed would be obvious, except for the widespread hope of finding a materialist explanation instead.[3] This book shows why that hope is pipe dream and why design is the most reasonable explanation for life's origin. Consider the following technologies to be found inside the cell:

- Information processing, storage, and retrieval.

- Artificial languages and their decoding systems.

- Error detection, correction, and proof-reading devices for quality control.

- Elegant feedback systems that monitor and regulate cellular processes.

- Digital data embedding technology.

- Signal transduction circuitry.

- Transportation and distribution systems.

- Automated parcel addressing ("zip codes" and "UPS labels").

- Assembly processes employing prefabrication and modular construction.

- Self-reproducing robotic manufacturing plants.

So far we have been considering eukaryotic cells, which are the sort that make up most of our bodies. They are the most complicated cells we know of. Some origin-of-life researchers might object at this point that, by focusing on such cells, we have set the bar too high. They would claim that eukaryotic cells needed to evolve in order to become this complex, and therefore these are not the first living forms that origin-of-life studies need to explain.

Implicit in this objection is the divide-and-conquer strategy of evolutionary biology. According to this strategy, to explain a complex system, one needs to explain how it could evolve from a simpler system. This strategy, in effect, merely restates Polya's dictum: solve hard problems by finding easy problems and solving them. Evolutionary biology, which presupposes the presence of life and attempts to explain its diversification, is fully committed to this strategy. But so are all materialist approaches to life's origin. They attempt to identify a likely sequence of chemical steps by which life may have formed under plausible prebiotic conditions.[4] Hence the preoccupation in origin-of-life research with "prebiotic" or "chemical" evolution.

Whatever its merits as a general principle for problem solving, the divide-and-conquer strategy has proven singularly ineffective in resolving the origin-of-life problem. It's true that eukaryotic cells are the most complicated cells we know. But the simplest life forms we know, the prokaryotic cells (such as bacteria, which lack a nucleus), are themselves immensely complex. Moreover, they are every bit as high-tech as the eukaryotic cells—if eukaryotes are like state-of-the-art laptop computers, then prokaryotes are like state-of-the-art cell phones. In particular, all the previous bullet points describing technologies found inside the cell apply also to prokaryotic cells.

Or, to revert to our previous analogy of an automated city, if an average eukaryotic cell is ten miles in diameter at a billion magnification, then an average prokaryotic cell is, on average, a mile in diameter. Thus, if eukaryotes are cities, prokaryotes are towns—yet they are towns that rival the technological sophistication of the larger cities. Not only do eukaryotic and prokaryotic cells perform all the basic functions associated with life (e.g., reproduction, growth, metabolism, homeostasis, well-defined internal organization, maintenance of boundaries, stimulus-response repertoire, and goal-directed interaction with the environment), but they do so by using many of the same basic structures.

For instance, the genetic code and the synthesis of proteins (by reading messenger RNA off DNA and then feeding it to ribosomes) is essentially the same in all cells, prokaryotic and eukaryotic. Ribosomes, the engines of protein synthesis, are themselves immensely complicated biochemical machines consisting of at least fifty separate proteins and RNA subunits. In fact, the very simplest prokaryotic cell requires hundreds of genes to handle its basic tasks of living.[5] Thus, even if it were possible to explain the origin of eukaryotic cells via the evolution of prokaryotic cells,[6] the origin of prokaryotic cells, which have essentially the same high-tech information processing capabilities of eukaryotic cells, would still need to be explained.

This level of complexity in even the simplest cells, however, raises a far more vexing problem for materialist origin-of-life research. Such research is committed to explaining the origin of life through a sequence of chemical steps that does not require intelligent input at any point. Standard dating places the origin of the Earth at about 4.5 billion years ago. For the first half billion or so years of its existence, the Earth was too hot and tempestuous for any life form. Then, at the moment when the Earth cooled sufficiently to permit life, *boom!* within a hundred or so million years, prokaryotic life appeared suddenly and abundantly (the best current estimates for its first appearance are 3.8 to 3.9 billion years ago).

While one hundred million years may seem like a long time in human terms, it is quite brief when measured against the age of the solar system and planets. One hundred million years ago, for example, was the middle Cretaceous period when dinosaurs and mammals roamed the Earth. While there have been many changes in life forms since then, no change as significant as the origin of life occurred. No change, that is, until the very recent and very sudden origin of the human mind (another phenomenon that is very poorly explained by materialist theories.)

There is no evidence whatsoever of earlier, more primitive life forms from which the early prokaryotes might have evolved. Origin-of-life researchers sometimes argue that later prokaryotic life forms were so successful that they consumed all the evidence of their evolutionary precursors. Yet the fact remains that we have no evidence of early life forms other than these. At a minimum, then, the problem of life's origin is to explain the origin of prokaryotic life and, in particular, its DNA-based protein synthesis apparatus. This is a huge unresolved problem. Philosopher of science Karl Popper explains why:

> What makes the origin of life and of the genetic code a disturbing riddle is this: the genetic code is without any biological function unless it is translated; that is, unless it leads to the synthesis of the proteins whose structure is laid down by the code. But, as [Jacques] Monod points out, the machinery by which the cell (at least the nonprimitive cell which is the only one we know) translates the code "consists of a least fifty macromolecular components *which are themselves coded in DNA.*" Thus the code cannot be translated except by using certain products of its translation. This constitutes a really baffling circle: a vicious circle, it seems for any attempt to form a model, or a theory, of the genesis of the genetic code.[7]

Popper wrote these words in the 1970s, but their impact is as powerful today. This book critiques the many attempts to resolve this problem.

2

A Batch of Red Herrings

Even though the origin of life is an entirely unsolved problem, you may be surprised to learn that origin-of-life researchers now prefer to talk about the origin*s* of life (plural) rather than the origin of life (singular). They argue that life must have originated in the universe a number of times. In their view, by focusing so much on the origin of Earth-based cellular life, we are discriminating against all the alternative life forms that exist elsewhere in the universe or predate cellular life on Earth. Perhaps some of these forms evolved into life on Earth. By focusing on life merely as we know it on Earth, we are being politically incorrect on a cosmic scale.

The umbrella discipline for advancing the claim that these alternative life forms exist is astrobiology. NASA, the principal funding source for astrobiology, defines this discipline as follows on its website:

> Astrobiology is the study of life in the universe. It investigates the origin, evolution, distribution, and future of life on Earth, and the search for life beyond Earth. Astrobiology addresses three fundamental questions: How does life begin and evolve? Is there life beyond Earth and how can we detect it? What is the future of life on Earth and in the universe?

The holy grail of astrobiology is "second genesis," which is the discovery, or invention in the lab, of an entirely new life form whose architecture and machinery is completely different from that of DNA-based cellular life.

In one sense, astrobiology is a perfectly valid discipline and the origins of life and second genesis is a perfectly valid pursuit. The problem, however, is that for now they apply to exactly one class of objects, namely, cellular life as we know it on Earth. Indeed, there is currently not a shred of evidence for any other form of life, whether here on Earth, on Mars, or elsewhere in the universe.

Astrobiology is an expanded version of exobiology. Exobiology covers all life outside of Earth (hence the "exo"), whereas astrobiology covers all life in the universe. With astrobiology, we have a population size of one: at least one place in the universe is known where life exists—Earth. With exobiology, we have a population of size zero: Earth is excluded and no other habitat for life in the universe is known. Exobiology is therefore a field of science without evidence, and astrobiology is a field of science whose evidence all derives from conventional biology.

In practice, the origins of life (plural), second genesis, and astrobiology are not so much scientific enterprises as exercises in hype. Until and unless there is solid evidence of full-fledged life forms (i.e., forms that reproduce, grow, metabolize, process information, interact with and adapt to changing environmental conditions, etc.) distinct from the DNA-based life here on Earth, the use of these terms is highly misleading. It suggests that we have scientific confirmation for these alternative life forms when in fact we have only the vain longing that they must be there. A prospector may likewise insist that "there's gold in them thar hills," but we do not call his quest a branch of the sciences. Throughout this book, therefore, we'll confine ourselves to the politically incorrect and cosmically unenlightened "origin of life" (singular).

Once life originates on planet Earth, evolutionary biologists assure us that it evolves. That raises a question: Is the ability to evolve part of the definition of life? Origin-of-life researchers now increasingly identify life with the ability to evolve. This is a significant change in definition. Many things can evolve without satisfying the more stringent conditions ordinarily demanded of living forms. Think, for example, of the evolution of a desert landscape, due to wind erosion over time. Its form evolves, but it is hardly alive. The shift in definition expands the concept of life to include a lot more than it should.

The minimal functional requirements for cellular life have traditionally been the following: reproduction, growth, metabolism, homeostasis, well-defined internal organization, maintenance of boundaries, stimulus-response repertoire, and goal-directed interaction with the environment. Notably absent from this list is evolution. By contrast, geophysicist and origin-of-life researcher Robert Hazen gives pride of place to evolution in his definition of life: "Most experts agree that life can be defined as a chemical phenomenon possessing three fundamental attributes: the ability to grow, the ability to reproduce, and the ability to evolve."

This definition is ill-conceived. Surely it is possible for living forms to vary within such strict limits that no evolution, in the sense of speciation, could occur. Asexual forms might reproduce so precisely (suppose, for instance, the copying mechanisms inside these cells were so exact as to rule out copying errors) that offspring were always identical to parents. Granted, such systems might have difficulty adapting to changing environments and thus be more likely to go extinct. But while they existed, they would be alive.

The ability to evolve is not a prerequisite for life but an additional property that living forms may or may not possess. We may reach the conclusion, after careful study, that a life form has the ability to evolve. Such a conclusion depends on case-by-case empirical evidence, bearing in mind that most historical species have simply gone extinct rather than continued to evolve. It cannot legitimately be reached by "definitional magic"—defining into existence whatever we wish to exist.

At this point, evolutionary biologists may attempt to justify evolvability as an essential feature of life by invoking the Law of Biogenesis. This law states that all life comes from life (in Latin, *omne vivum ex vivo*). In our ordinary experience, this law appears to hold without exception. Nevertheless, it faces an obvious exception in the origin of life: since life has not always existed, there must come a point (or points) at which life arose from nonlife.

Accordingly, the Law of Biogenesis does not hold universally but only *after* life originates (regardless of whether it originates by strictly material mechanisms or by intelligent design). Given that this law must be suspended whenever life arises from nonlife, one may ask whether in the history of life it was suspended only once or more than once. In evolutionary terms, this is to ask whether the history of life can be represented as a single tree (i.e., universal common ancestry) or as multiple trees.

By itself, the Law of Biogenesis says nothing about the number of starting points for life or about life's subsequent evolvability. That requires an independent assumption. Many materialist biologists, by regarding it as highly unlikely that the same genetic code could originate more than once, assume that life began only once and therefore had to be highly evolvable in order to produce the diversity that we see. But a single origin of life on Earth is becoming increasingly controversial. For instance, molecular evolutionists Carl Woese and W. Ford Doolittle now argue for multiple origins of life. In any case, one cannot derive universal common ancestry, and therefore a requirement that life forms be able to evolve, from the Law of Biogenesis.

Origins of life (plural), astrobiology, exobiology, second genesis, and the demand that evolvability be an essential feature of life are red herrings. They distract origin-of-life researchers from the magnitude of the task before them, which is to explain the origin of cellular life on Earth, the only life that we in fact know and for which we have evidence.

3

Spontaneous Generation

Science is commonly thought to be a cumulative enterprise that relentlessly pushes back the frontiers of knowledge. This image of science is mistaken. Revolutions and retractions are also an integral part of science.[1] Science is an interconnected web of theoretical and factual claims about the world that are constantly being revised in light of new evidence and for which changes in one portion of the web can induce radical changes in another. In particular, science regularly confronts the problem of having to retract claims that it once confidently asserted.[2]

One such claim concerns the regular formation of life from nonlife in everyday circumstances. It seems far-fetched to us now, but as late as the nineteenth century people believed that full-fledged animals could originate suddenly, without parents, from mud, rags, or decaying organic matter such as rotting meat. Today the idea appears no more than a superstition, but at one time both observation and common sense seemed to confirm it. Leave dirty rags in the corner of a shed, and doesn't it soon become a nest of mice? Leave rotting meat out, and isn't it quickly covered with maggots? Decaying meat seemed always to be covered with swarming flies. As a consequence, the flies were assumed to have originated from it. Thus, it was widely believed that animals could arise on their own, full-blown, from nonliving matter. The belief was called *spontaneous generation.*

The rise of modern science resulted in concerted efforts to find out exactly how things happened. Under systematic investigation, belief in spontaneous generation began to wane. In 1668 Francesco Redi conducted an experiment to determine whether worms arose spontaneously in decaying food. He placed similar samples of raw meat in two sets of jars. One set he covered with a muslin screen, the other he left open. After several days, the muslin screen covering the first sample was sprinkled with fly eggs, but there were none on the meat itself. The meat in the open jar was covered with eggs, which soon hatched into maggots. Redi had shown that maggots were not simply small worms that arose spontaneously but rather were fly larvae. Redi's experiment cast doubt on the spontaneous generation of macroscopic organisms (i.e., organisms large enough to be visible to the naked eye).

After the invention of the microscope, scientists could observe bacteria. These single-celled organisms seemed to be the simplest living forms. Could they have originated spontaneously from nonliving chemicals? Even if spontaneous generation could no longer explain the appearance of multicelled organisms, perhaps it could explain the appearance of much simpler single-celled organisms. And what if these were the first living forms on Earth?

Such speculations matched up nicely with Darwin's *Origin of Species* (1859). In that book Darwin hypothesized how species might evolve from already existing species. Darwin's theory purported to explain how life could have become gradually more complex starting from one or a few simple forms. Nevertheless, it did not explain, nor did it attempt to explain, how life had arisen in the first place.

Darwin speculated on the origin of life in only one place—an unpublished letter written in 1871 to Joseph Hooker. In that letter he sketched how life might have originated through a series of chemical reactions:

> It is often said that all the conditions for the first production of a living organism are now present, which could ever have been present. But if (and oh! what a big if!) we could conceive in some warm little pond, with all sorts of ammo-

nia and phosphoric salts, lights, heat, electricity, etc. present, that a proteine [*sic*] compound was chemically formed ready to undergo still more complex changes, at the present day such matter would be instantly devoured or absorbed, which would not have been the case before living creatures were formed.[3]

In the 1870s and 1880s evolutionary thinkers such as Ernst Haeckel and Thomas Henry Huxley also began to speculate about how life originated. Haeckel and Huxley thought that the problem of resolving life's origin would be fairly simple because they assumed that life was in essence a chemically simple substance, which they called "protoplasm." Thus, according to Haeckel, the cell was an enclosed blob of jelly or, as he called it, a "homogeneous globule of plasm."[4] Both Haeckel and Huxley thought protoplasm could be easily constructed by combining and recombining simple chemicals such as carbon dioxide, oxygen, and nitrogen. Huxley even blundered famously by positing mud dredged up from the bottom of the ocean, which he called *Bathybius haeckelii* in honor of his friend Haeckel, as the primordial stuff from which life arose. Soon thereafter this mud was found to be an inorganic precipitate lacking all the distinctive qualities normally associated with life.[5]

Over the next sixty years biologists and biochemists gradually revised their view on the nature of life. Origin of life theories of the time reflected this increasing awareness of cellular complexity. Nineteenth-century theories of life's origin envisioned life as arising almost instantaneously via a one- or two-step chemical process. In effect, they proposed that the simplest life forms were capable of forming by spontaneous generation. Yet, by the twentieth century it became evident that not even the (comparatively) simplest life forms could be produced by spontaneous generation. By the 1930s most biologists had come to see the cell as a complex metabolic system.

In the nineteenth century, the last outpost of spontaneous

generation as an everyday event in nature was the world under the microscope. Microscopic creatures were so small and appeared to be so simple that it was not difficult to believe they arose spontaneously from nonliving matter. After all, if bits of straw were left to rot in a pan of water, the water was soon swarming with bacteria. And bacteria were, according to the science of the day, just blobs of protoplasm.

Notwithstanding, Louis Pasteur showed that even here spontaneous generation failed. In the early 1860s, two centuries after Francesco Redi had produced convincing evidence against the spontaneous generation of macroscopic organisms, Pasteur produced equally compelling evidence against the spontaneous generation of microscopic organisms. In a famous set of experiments, Pasteur showed that water could be kept free of bacteria by boiling it and then exposing it only to purified air. By doing so, he demonstrated that the microscopic bacteria that mysteriously rotted the straw in water had in fact arrived on air currents. Pasteur's elegant experiments showed that the growth of microbes in otherwise sterile media was due to contamination by existing microbes, not to spontaneous generation of new ones.

Pasteur's experiments reversed the onus of proof. Those who argued for spontaneous generation needed to demonstrate an unambiguous example. It was left to physicist William Tyndall to administer the *coup de grace* to spontaneous generation. Tyndall thought it odd that people with broken ribs that had pierced their lungs but not their skin (i.e., the damage was purely internal) did not develop infections even though they continued to breathe nonsterile air. In 1876 Tyndall therefore decided

> to replicate the sterilizing effects of lungs in the laboratory, employing a simple device that became known as a Tyndall box. He coated the insides of a black-painted box with a thin layer of sticky glycerin. When the box was put in a quiet place, all the floating particles inside it soon settled out or collided with the sides and became trapped. The air inside the box could be seen to be completely transparent when a

beam of light was shone through it. At this point, sterile solutions of any sort that were exposed inside the box would remain sterile indefinitely. Tyndall boxes were put on display at the Royal Society in London, where they convinced everyone who saw them. . . . The battle of Tyndall's boxes was the last skirmish in a war that had lasted three hundred years, from the time of Francesco Redi. . . . By quite literally settling the dust of this final skirmish, Tyndall managed to resolve the matter.[6]

Tyndall boxes disposed of an idea that had been held, in one form or another, for thousands of years. Thereafter scientists rejected spontaneous generation as an explanation for the abrupt appearance of microscopic organisms.

In the decades following Pasteur and Tyndall, science's understanding of the complexity of cells greatly increased. Advances in cell biology and biochemistry during the rest of the nineteenth and early years of the twentieth century provided additional reasons for ruling out the routine abrupt origin of life from nonlife. During this period, the view that life comes only from preexisting life was universally accepted. Nearly every scientist agreed that cells come only from cells and that even the simplest cell was not generated spontaneously. The idea of spontaneous generation appeared all but dead.

Yet if Darwinian evolution were correct, complex forms of life ought to have evolved by materialist processes from simpler ancestors. Redi, Pasteur, and Tyndall had shown that full-blown organisms—whether mice, maggots, or microbes—do not arise from nonliving matter. Nevertheless, Darwinism, the dominant view of evolution, seemed to point to a purely materialist origin for life. To be sure, Darwin's actual theory focused on the formation of new organisms from existing organisms by purely material processes. But the question whether the origin of life might come about by purely material processes was not far behind. Indeed, without a materialist explanation of life's origin, Darwin's explanation of the origin of species remains fundamentally incomplete.[7]

Scientists therefore continued to search for a materialist

explanation of life's origin. During the first two decades of the twentieth century, many advances were made in the study of viruses and the chemistry of living matter. For instance, colloid chemistry became an important field of study during this period. Colloids are particles that make up gels. The size of many colloids is approximately that of large cells, and colloids seem to share some features with cells. Colloids were therefore thought to offer insight into the origin of life. At the same time, amino acids and other simple building blocks of cells were synthesized in the laboratory, much as sugars had been synthesized in the nineteenth century. Through these and other studies, scientists gained increasing insight into the chemical makeup of cells. Origin-of-life research thus became a program for showing how the chemical building blocks of life could have originated and organized themselves into living forms by purely material processes.

4

Oparin's Hypothesis

In 1924, Russian biochemist Alexander Oparin proposed a purely materialist approach to life's origin that has set the tone for origin-of-life research ever since.[1] According to him, the first cell originated from nonliving matter. That wasn't new, of course; his new idea was that it didn't happen all at once. Instead, life arose gradually in stages.[2] Oparin argued that simple chemicals combined to form organic compounds, such as amino acids. These in turn combined to form large, complex molecules, such as proteins. And these aggregated to form an interconnecting network inside a cell membrane.

According to Oparin, the atmosphere of the early Earth was very different from the present one. Energy sources such as lightning or heat from volcanoes were said to act on simple carbon compounds in the atmosphere, transforming them into more complex compounds. In the Earth's early seas these newly formed compounds were said to come together to form microscopic clumps, the forerunners of the first living cells on Earth.

In 1928 English biochemist J. B. S. Haldane put forward essentially the same idea.[3] He thought that ultraviolet light from the sun caused simple gases in the Earth's primitive atmosphere (e.g., hydrogen, methane, water vapor, and ammonia) to transform into organic compounds, turning the primitive ocean into a hot, dilute "soup." Out of this soup came virus-like particles that eventually evolved into the first cells.

Oparin and Haldane thus laid the groundwork for a theory of *prebiotic* or *chemical evolution,* according to which life began in a sea of chemicals, sometimes called the *prebiotic soup*. To this day, their view is known as the Oparin hypothesis or the Oparin-Haldane hypothesis. Despite many modifications to the hypothesis in later years, it exemplifies the standard contemporary evolutionary approach to the origin of life.

According to the Oparin hypothesis, chance by itself could not drive the interactions of chemicals and compounds needed to form complex biomolecules and from there life. Instead, the hypothesis posits that some internal tendency of matter toward self-organization gives rise to the ordered structures we see in life. Let us now examine this hypothesis in more detail, noting carefully the assumptions on which it is built.

Assumption #1: Reducing Atmosphere. *The Earth's early atmosphere contained abundant free hydrogen and little or no oxygen.*

According to Oparin, the first cells arose gradually over millions of years. He believed that conditions at the surface of the early Earth allowed a massive accumulation of organic compounds before life began. Such an accumulation would improve the probability that all the right compounds could come together at some point and combine into a cell.

This accumulation of organic compounds could not have occurred, however, if the Earth's atmosphere contained significant amounts of free oxygen (O_2). Oxygen destroys organic compounds by reacting with them in a process called *oxidation*. Oparin believed that the early atmosphere must have been composed of hydrogen (H_2) and hydrogen-rich gases such as methane (CH_4), ammonia (NH_3), and water vapor (H_2O), but not oxygen. Such an atmosphere is called a *reducing atmosphere*.

Furthermore, Oparin believed that the first cells were *anaerobic* (able to survive without oxygen) and *heterotrophic* (unable to make their own food), obtaining many of their essential nutrients instead from the surrounding water. These anaerobic heterotrophs were thus said to obtain their energy by fermentation, a method of releasing energy from organic molecules in the absence of free oxygen.

How did Oparin justify a reducing atmosphere? Oparin argued that because hydrogen is the most common element in the universe, early in the history of the universe it would readily have combined with other light elements to form hydrogen-based compounds such as methane and water. All the free oxygen would therefore have been used up.

Assumption #2: Preservation. *Simple organic compounds that formed in the primeval soup were somehow preserved so that the energy responsible for their formation did not also destroy them.*

Atmospheric gases could have reacted to form more complex compounds only if energy was available to cause them to react. Ultraviolet light from the sun, cosmic rays, electrical energy from lightning bolts, heat, and radioactivity might have provided the necessary energy. According to Oparin's hypothesis, the available energy converted the atmospheric gases into more complex compounds such as sugars, amino acids, and fatty acids. But the same energy sources could also have broken down the molecular structure of these compounds. Oparin therefore assumed that these compounds were somehow protected from such destructive effects and collected in the Earth's primitive oceans to form a soup out of which life could emerge.

Assumption #3: Concentration. *Biological compounds accumulated in sufficiently high concentration so that they could combine with each other to form the large, complex molecules needed for life.*

Oparin proposed that, as simple organic compounds accumulated in the primitive oceans, they were not so diluted as to exclude biologically significant interactions. Rather, material forces were able to set aside and concentrate these compounds, thereby enabling them to combine with each other to form more complex substances such as proteins, nucleic acids (RNA and DNA, whose double-helix structure was unknown to Oparin when he proposed his hypothesis), polysaccharides (long chains of sugar molecules), and lipids (fats).

19

At first, these large biological molecules were far simpler than their biochemical counterparts today, but gradually they became more and more complex. Eventually, catalytic proteins with the ability to accelerate reaction rates emerged. They were the forerunners of the first enzymes. In light of this assumption, scientists came to believe that amino acids reacted with each other to form larger amino-acid chains (polypeptides), that nucleotide bases combined with each other to form long DNA and RNA sequences, and that simple sugars combined with other simple sugars to form complex sugars (polysaccharides).

Assumption #4: Uniform Orientation. *Only "left-handed" or L-amino acids combined to produce the proteins of life, and only "right-handed" or D-sugars reacted to produce polysaccharides and nucleotides.*

Some time early in their history, cells developed a preference for only one orientation of amino acid and only one orientation of sugar. Two amino acids can be chemically identical, which means that they have the same chemical structure. Nevertheless, they can differ in their three-dimensional shape in the same way that the left hand differs from the right. In other words, they can be mirror images of each other. Proteins that make up living things consist only of left-handed amino acids (except glycine, which is neither right- nor left-handed). Yet, in the absence of life, left- and right-handed forms are equally probable in nature. Sugars also occur in mirror images. The genetic material of living organisms (DNA and RNA) is composed of nucleotide bases along a sugar-phosphate backbone. These sugars are all right-handed. Here again, in the absence of life, left- and right-handed forms are equally probable in nature.

Assumption #5: Simultaneous Emergence of Interdependent Biomacromolecules (such as DNA and proteins). *The genetic machinery that tells the cell how to produce proteins and the proteins required to build that genetic machinery both origi-*

nated gradually and were present and functioning in the first reproducing protocells.

In living cells today, DNA and protein depend on each other for their existence. Scientists differ about how one could have originated without the other. Some maintain that DNA's genetic coding came first. Others maintain that functioning proteins came first. Still others maintain that both DNA and protein appeared simultaneously. In recent years the "RNA first" view has gained prominence. Also known as "ribozyme engineering," this view holds that RNA with catalytic properties was the precursor to both DNA and proteins. All such proposals for the materialist origin of these biomacromolecules are highly speculative.

Assumption #6: Functional Integration. *The highly organized arrangement of thousands of parts in the cell's chemical machinery, which is needed to achieve the cell's specialized functions, originated gradually in coacervates or other protocells.*

In present-day life, biological macromolecules (such as DNA and proteins) are parts of a much more complicated functionally integrated system—the cell. According to Oparin, before the origin of cells, biological macromolecules combined to form complex microscopic aggregates called coacervates. Coacervates are organized droplets of proteins, carbohydrates, and other materials formed in a solution. They may have "competed" with one another for dwindling supplies of "food" molecules in the primitive oceans and thus been the forerunners of the first living cells. Oparin thought of this competition as a form of Darwinian natural selection resulting in the survival and domination of ever more complex and lifelike coacervates until a true cell finally appeared. These first cells would have had cell membranes, complex metabolism, genetic coding, and the ability to reproduce; moreover, they would have dominated the primitive seas.

Assumption #7: Photosynthesis. *The chemical process known as photosynthesis, which captures, stores, and utilizes the en-*

ergy of sunlight to make nutrients, gradually developed within coacervates.

Oparin speculated that the further development of primitive heterotrophic organisms (organisms unable to make their own food from inorganic starting materials) resulted in the formation of cells capable of photosynthesis. These were the first autotrophic organisms (organisms able to make their own food). According to Oparin, the driving force for this evolutionary development was the gradual decrease of organic nutrients from the primitive oceans (such molecules having been largely consumed by the heterotrophic organisms).

Oparin therefore proposed that spontaneous chemical events within the coacervates led to the formation of photosynthesis. Photosynthesis captures and processes energy from sunlight. According to Oparin, photosynthesis supplied the energy needs of primitive cells. Since photosynthesis releases free oxygen (O_2) into the air, Oparin argued that free oxygen was not available on the Earth until after the emergence of photosynthesis. Only then could autotrophic organisms develop, which used oxygen in respiration and thus did not have to depend on fermentation to supply organic nutrients.

In summary, Oparin visualized a gradual origin of life from nonliving organic chemicals by a long process spread over hundreds of millions of years and without the aid of intelligent agency. Instead of cells appearing suddenly through spontaneous generation (as was still believed in Darwin's day—see chapter 3), the Oparin hypothesis breaks the origin of life down into seven stages. Moreover, because transitions from one stage to the next could, at least to some extent, be tested experimentally, the hypothesis itself could be tested.

Thus, Oparin's hypothesis was more than just a revival of spontaneous generation, the idea that full-blown cellular life can emerge abruptly from nonlife. Rather, it employed both a divide-and-conquer and a self-organizational strategy to explain the cell's complexity. It used divide-and-conquer to break the problem of life's emergence into seemingly manageable steps. Moreover, at each step it appealed to self-organizational properties of matter to bring about the next needed advance in the progression from nonlife to life.

We critique the seven assumptions that make up the Oparin hypothesis in chapters 7 to 13.

5

The Miller-Urey Experiment

One of the advantages of Oparin's hypothesis was that it could to some extent be tested. Not directly tested, of course, because we cannot observe a past event such as the origin of life. But scientists can construct hypothetical scenarios for events that might have happened and then set up laboratory experiments to see if similar events could occur today. These are called *simulation experiments.* They are designed to simulate what might have happened on the early Earth when life began. Results from such simulation experiments could then assess the reasonableness of Oparin's hypothesis. Although the hypothesis itself was sketchy and relied a great deal on convenient assumptions, Oparin and his followers were confident that it was only a matter of time before laboratory experiments would fill in the details of how life first developed.

Oparin proposed that life arose from chemical reactions among simple gases in the atmosphere: methane, ammonia, hydrogen, and water vapor. These reactions would be activated by various forms of energy present on the Earth before life began—by lightning, heat from volcanoes, kinetic energy from earthquakes, and light from the sun. When atmospheric gases encountered this energy, they would react to form organic compounds such as amino acids, fatty acids, and sugars.

How has this hypothesis been tested in the laboratory? Scientists have taken the simple gases suggested by Oparin, enclosed and mixed them together in a glass apparatus, and then subjected them to various energy sources, such as ultraviolet light (to simulate sunlight) and electrical discharges (to simulate lightning). Such experiments are called *primitive atmosphere simulation experiments,* and many have been performed since the early 1950s. These experiments attempt to reproduce likely conditions on the early Earth. As such, they cannot provide direct observation of life's origin. The purpose of these experiments is to determine what compounds might reasonably have been formed on the primitive Earth prior to the emergence of life and whether any of these compounds play a role in life.

In 1953, Stanley Miller and Harold Urey reported the first such experiment. At the time, Miller was a graduate student at the University of Chicago working with Urey (who had won the Nobel Prize in chemistry in 1934). When Miller began his graduate work, nobody had yet carried out experiments to see if the primitive atmosphere assumed by Oparin would indeed produce organic compounds necessary for life. Miller and Urey were interested in doing just that.

To simulate the conditions that Oparin assumed to exist on the primitive Earth, Miller and Urey designed the following experiment. They boiled water in a round-bottomed flask, thereby saturating the atmosphere of the flask with water vapor. They then eliminated any trace of O_2. Next they piped in methane and hydrogen gases, and then they generated ammonia gas by heating dissolved ammonium hydroxide (NH_4OH) in the water at the bottom of the flask. During the experiment, the water in the flask boiled, driving the gases in a clockwise direction through the apparatus. At the top of the apparatus was a 5-liter glass sphere containing two electrodes connected to a source of electricity. As the gases passed between these electrodes, they were subjected to 50,000-volt sparks.

Leaving the spark chamber, the gases passed through a cooling device that condensed the water vapor and any non-volatile (nongaseous) organic compounds that formed in the glass sphere. This solution then collected in a trap at the bottom of the apparatus. Miller analyzed the gooey, tar-like substance formed in the flask and identified several of the amino acids found in proteins today, including glycine, alanine, aspartic acid, and glutamic acid. He also found several nonbiological amino acids, as well as urea and some simple organic compounds such as formic acid, acetic acid, and lactic acid.

Since Miller's early work in the 1950s, other biological compounds have been detected in similar atmospheric simulation experiments. The list now includes many essential organic compounds found in living things. Imagine the excitement in the scientific community as the results of these experiments were first published. The possibility of humans creating life in the laboratory seemed just around the corner. These experiments seemed to prove that many of the chemical building blocks of life could have formed naturally under conditions assumed to have existed on the primitive Earth (one of the main assumptions being a reducing atmosphere).

Experimental evidence thus seemed to support the first stage in Oparin's hypothesis. In consequence, Oparin's views on chemical evolution gained credibility and new adherents. But when scientists sought to go beyond the simplest building blocks of life, they were quickly frustrated. The step from simple compounds to the complex molecules of life, such as proteins and DNA, has proved exceedingly difficult. Thus far, this step has resisted all efforts by scientists working on the problem.

The reason for this failure is that the needed chemical reactions do not occur. As long as no oxygen is present, the reactions that produce the simple building blocks of life occur readily in the laboratory, but the chemical reactions required to form proteins and DNA do not. In fact, these biomacromolecules haven't been produced in any simulation experiment to date. In addition, the assumptions underlying primitive atmosphere simulation experiments have proven problematic. Such experiments should, after all, simulate realistic prebiotic conditions (i.e., conditions that might reasonably be expected on the early Earth). Yet many do not.

6

Primitive Undersea Simulation Experiments

Our best evidence for the Earth's early atmosphere suggests that it did not favor the origin of life because it had free oxygen, which interferes with the reactions needed for life. Cells that scientists study today have complex systems that manage the flow of oxygen, but it is not clear how alleged precursors to cells could have such systems. Some researchers, therefore, speculate that life originated at undersea hydrothermal vents where oxidation may have played less of a role in hindering the formation of life's building blocks.

Oceanographer Jack Corliss discovered hydrothermal vents in 1977. Despite the intense heat released by these vents from the Earth's interior, life abounds around them. Corliss hypothesized that this is where life might have first arisen. His proposal runs counter to earlier speculation, such as that of Oparin, which took the sun as the crucial energy source driving the origin of life. For Corliss, heat from the Earth's interior and high pressure from the ocean depths provided the crucial energy driving life's origin.[1]

Corliss's proposal has attracted much attention. Organic chemist Günter Wächtershäuser has tried to fill in the details, speculating that metal sulfide minerals such as iron pyrite, which are abundant in seafloor rocks, might catalyze the chemical reactions needed to form biologically relevant compounds (a task that living cells today perform with enzymes).[2] Researchers at the Carnegie Institution have been testing Corliss's proposal through high-pressure, high-temperature simulation experiments that attempt to recreate ocean conditions at the hydrothermal vents.[3]

As with atmospheric simulation experiments, researchers have found that biologically relevant compounds (including amino acids) can be formed in such undersea simulation experiments. To prevent such compounds from degrading, however, these experiments control for the effects of free oxygen. Yet in doing so, they may not be simulating authentic prebiotic conditions.

If oxygen was present in the Earth's primitive atmosphere, then seawater synthesis experiments do not redress the problem of oxidation of organic compounds *by atmospheric oxygen*. Current oceans are in equilibrium with the atmosphere, and seawater contains enough oxygen (O_2) to oxidize the current levels of dissolved organic compounds many times over. The only reason we find dissolved organic compounds in seawater today is that biological processes constantly add to them. Stop the supply, and oxygen will disintegrate the existing organic compounds. There is no good reason to think that the situation was different on the early Earth if oxygen was present. Sea water does not protect organic compounds from decomposition by oxygen.

All the problems associated with atmospheric simulation experiments therefore remain. In particular, high-pressure, high-temperature undersea simulation experiments do nothing to address the fundamental problem that plagues all origin-of-life scenarios, namely, how to arrange the basic building blocks of life (amino acids, nucleic acids, lipids, etc.) into the highly organized, information-rich structures required for life.

Undersea simulation experiments also introduce a new problem. Water at high pressure and high temperature, especially if shooting out of vents, is turbulent. Even with the building blocks for life in place, the turbulence of such watery environments will hinder the further organization of these building blocks into biomacromolecules. Yes, the origin of life requires energy. But this energy needs to be directed, and the haphazard motion of water at high temperature and high pressure impedes such direction.[4]

Primitive undersea simulation experiments therefore create more difficulties than they resolve.

Free Oxygen

In the next seven chapters we critique the seven assumptions that make up the Oparin hypothesis (see chapter 4). Let's begin with Oparin's assumption that the early Earth's atmosphere was devoid of oxygen. All experiments simulating the atmosphere of the early Earth have excluded free oxygen (i.e., oxygen in a dissolved or gas state, not bound up with other substances). That's because oxygen acts like a wrench in a gear-box, actively hindering the chemical reactions that produce organic compounds. Moreover, if any such compounds did happen to form, free oxygen would quickly destroy them in a process called oxidation. That's why many food preservatives are antioxidants—they protect food from the effects of oxidation.

As a consequence, the standard story of chemical evolution assumes that there was no oxygen present in the Earth's atmosphere at the origin of life. Yet scientists now have strong geological evidence that significant amounts of oxygen were present in the Earth's atmosphere from the earliest ages (and thus at the time the first life was supposed to be forming).[1] For instance, many minerals, such as rusting iron, react with oxygen. The resulting oxides are found in rocks dated earlier than the origin of life.

If oxygen had been present in the Earth's early atmosphere, organic compounds could not have formed and accumulated the way they did in the Miller-Urey experiment. Such experiments also leave a paradox unresolved. Oxygen would have prevented the accumulation of organic compounds on the early Earth. Yet without oxygen, as Oparin and the Miller-

Urey experiment assumed, organic compounds may not have accumulated either. Significant levels of oxygen would have been necessary to produce ozone. Ozone shields the Earth from levels of ultraviolet radiation lethal to biological life. Since life did in fact flourish on the early Earth, a realistic simulation of the early Earth's atmosphere may need to include oxygen.

Another problem is that the Oparin hypothesis and the Miller-Urey experiment assumed that the Earth's early atmosphere was hydrogen-rich. Yet geochemists concluded in the 1960s that the Earth's primitive atmosphere was derived from volcanic outgassing, and consisted primarily of water vapor, carbon dioxide, nitrogen, and only trace amounts of hydrogen. Because the Earth's gravity is too weak to retain light hydrogen gas in the atmosphere, most of the volcanic hydrogen would have been lost to space. With no hydrogen to react with the carbon dioxide and nitrogen, methane and ammonia could not have been major constituents of the early atmosphere.

Geophysicist Philip Abelson concluded in 1966: "What is the evidence for a primitive methane-ammonia atmosphere on Earth? The answer is that there is no evidence for it, but much against it."[2] In 1975, Belgian biochemist Marcel Florkin announced that "the concept of a reducing primitive atmosphere has been abandoned," and the Miller-Urey experiment is "not now considered geologically adequate."[3] Sidney Fox and Klaus Dose conceded in 1977 that a reducing atmosphere did "not seem to be geologically realistic because evidence indicates that . . . most of the free hydrogen probably had disappeared into outer space and what was left of methane and ammonia was oxidized."[4] Since 1977 this view has become a near-consensus among geochemists. As Jon Cohen wrote in *Science* in 1995, many origin-of-life researchers now dismiss the 1953 experiment because "the early atmosphere looked nothing like the Miller-Urey simulation."[5]

What if the Miller-Urey experiment is repeated with a more realistic mixture of water vapor, carbon dioxide, and nitrogen? Fox and Dose reported in 1977, and Heinrich Holland reiterated in 1984, that no amino acids are produced by sparking such a mixture.[6] In 1983, Miller reported that he and a colleague were able to produce a small amount of the simplest amino acid, glycine, by sparking an atmosphere containing carbon monoxide and carbon dioxide instead of methane as long as substantial free hydrogen was present.[7] But he conceded that glycine was about the best they could do in the absence of methane. John Horgan summarized the state of this research for *Scientific American* in 1991: an atmosphere of carbon dioxide, nitrogen and water vapor "would not have been conducive to the synthesis of amino acids."[8]

Oparin's assumption that the early Earth had a reducing atmosphere conducive to the pre-biotic synthesis of amino acids (Assumption #1 in chapter 4) appears to be incorrect.

8

Reversible Reactions

An obstacle faced by any theory of chemical evolution can be stated as a paradox. Some chemicals react quite readily with one another. They connect easily, like the north and south poles of two magnets. Others resist reacting. Getting them to react is like forcing the magnets' north ends together. To drive such a chemical reaction forward requires energy (heat, for example, or electricity). But—and here is the paradox—energy also breaks chemical compounds apart.

Energy is therefore a double-edged sword, building up complex molecules from simpler parts but also breaking down developing molecules. In the formation of life, it is therefore crucial that the destructive effects of energy not outpace the constructive effects. For life to form, there has to be a proper balance, or equilibrium, between these two opposing tendencies, the tendency for energy to build things up and its tendency to break things down. Yet, when we take into account the destructive effects of energy on the early Earth, the equilibrium state of the "primeval soup" would favor not complex molecules but simple ones that have no inherent capacity for spontaneously organizing themselves into the machinery of a living cell.

Even the preservation of simple organic molecules on the early Earth is not as straightforward as one might think. Origin-of-life researchers are quick to note that the Murchison meteorite carried numerous organic compounds to Earth. Thus, it is thought, if chemical pathways to organic compounds fail on the early Earth, organic matter can always hitch a ride on meteorites to get here. But, can enough organic matter, and of the right sort, be supplied to Earth in this way to jump-start life (it's not as though meteorites make scheduled deliveries of correct components—unless the deliveries were themselves intelligently designed)? As we shall see in the next point, organic matter produced under plausible prebiotic conditions tends to be useless for building up biological complexity.

In any case, the evidence for the early Earth as a preserver of simple organic molecules is hardly compelling. Atmospheric simulation experiments, such as the one by Miller and Urey, neglect the destructive effects of energy. In such experiments, amino acids and other organically relevant products that form are siphoned off through a trap to protect them from breaking down. In the trap, they are preserved from the destructive effects of the electrical discharges. But suppose the amino acids and other products had been continually exposed to such discharges, as on the early Earth. In that case, they would have disintegrated as soon as they formed, and Miller could not have detected them with his apparatus. We have no evidence that any such protective mechanism existed on the early Earth.

Oparin's assumption that simple organic compounds were somehow preserved on the early Earth (Assumption #2 in chapter 4), though possible, lacks a mechanism, and is thus without evidence.

Interfering Cross-Reactions

Many reactions needed to form biologically important compounds have been observed under (and often *only* under) artificial laboratory conditions. At the same time, many reactions that occur in nature work against the formation of biologically important compounds. Amino acids, for instance, do not readily react with each other. They do, however, readily react with other substances, such as sugars. But this creates a problem. If amino acids formed on the early Earth, they would not float around in lakes and ponds simply waiting for the correct amino-acid partners to show up to form proteins. Instead, they would combine with other compounds in all sorts of cross-reactions, tied up and unavailable for any biologically useful function. This explains why, in prebiotic simulation experiments, the predominant outcome is large yields of nonbiological sludge.[1]

For Oparin's account of life's origin to hold up, simple biological compounds would need to be concentrated in sufficient quantities before they could combine to form the complex biomacromolecules necessary for life. This is Oparin's assumption of "concentration." But there is no evidence that nature can be concentrated simple biological compounds so that they later arrange themselves into the large complex molecules needed for life. Nature provides no lay-away plan for setting aside biologically important compounds for future use.

In an organic soup full of diverse chemicals, amino acids would not tend to combine with amino acids or sugars with sugars. Rather, interfering cross-reactions would tie up biologically relevant chemical compounds and render them biologically useless.[2] Primitive atmosphere simulation experiments confirm this point. No biopolymers (chains of amino acids or nucleotides) useful to life have been found in such experiments except for some very small peptides. Most of what is produced in such experiments is nonbiological sludge—hydrogen-poor, insoluble materials known as tars.[3] This shows that interfering cross-reactions occur under even the most favorable experimental conditions, casting real doubt on Oparin's hypothesis of concentration.

Further doubt arises because there is no geological evidence of any significant prebiotic accumulation of organic matter. It is often claimed that all such evidence disappeared as soon as life arose and organisms began to feed on that matter. But abundant clay deposits from the time of life's origin have been found. They should retain large amounts of hydrocarbons and nitrogen-rich compounds from the prebiotic soup if they had been present. The surface of the clay has tiny cavities that could imprison these molecules, where they would still be evident today. Thus, if the prebiotic soup had really existed, we should expect to find surviving traces of it in the oldest rocks. In fact, we don't find any such traces.[4]

Oparin's assumption that simple organic compounds were concentrated in sufficient quantities to form the biomacromolecules needed for cellular life (Assumption #3 in chapter 4) therefore appears to be incorrect.

10

Racemic Mixtures

Amino acids, sugars, proteins, and DNA are not simply bundles of chemicals. They exhibit very specific three-dimensional structures. When synthesized in the laboratory, they may have the right chemical constituents but still exhibit the wrong three-dimensional form. For example, amino acids appear in two forms or *chiralities*. These are mirror images of each other just as a right glove is the mirror image of a left glove. The two forms are referred to as right- and left-handed amino acids. Living things use only left-handed amino acids (L-amino acids) in their proteins (and thus are said to be *homochiral*). Right-handed ones (R-amino acids) don't "fit" the metabolism of the cell any more than a right-handed glove fits a left hand. If just one right-handed amino acid finds its way into a protein, the protein's ability to function is diminished and often destroyed.

Although the amino acids that make up the proteins of living cells are all "left-handed" (the L-form), in simulation experiments such as the one by Miller and Urey, the amino-acid products found in their apparatus are always a racemic mixture, that is, 50 percent left-handed and 50 percent right-handed. No one knows why amino acids in living things only occur in one of their two possible forms. Yet if purely material processes can originate life, they must be capable of concentrating the L-forms of amino acids in specific locations.

Is there any evidence that material processes can concentrate only the correct forms of amino acids necessary for life? Perhaps the closest thing to evidence is work by geochemist

Robert Hazen and fellow researchers. They placed four calcite crystals in a racemic mixture of the amino acid aspartic
acid. Two of the crystals were smooth, and two had microscopic terraces. The researchers found that one of the terraced
crystals had a slight excess of right-handed amino acids and
that the other had a similar excess of left-handed amino acids.
The most extreme disproportion that they found was 55–45
instead of the usual 50–50. By contrast, the smooth crystals
did not distinguish between the L- and D-forms of aspartic
acid.[1]

This is hardly compelling evidence that purely material
processes can segregate L-amino acids from D-amino acids
with the degree of precision required for life. Hazen's research
applies to only one amino acid—aspartic acid. Moreover, his
calcite crystals merely induced a slight disproportion in the
ratio of L- to D-amino acids. Life as we know it requires *pure
concentrations* of the L-form of amino acids—a substantial
admixture of the D-form, as in Hazen's experiments, will not
do. As a consequence, Noam Lahav of the Hebrew University
in Jerusalem does not regard Hazen's research as providing a
general mechanism by which L-amino acids can be segregated
from R-amino acids.[2]

Like amino acids, sugars come in two forms (or chiralities)
but only appear in one form in living things. When sugars are
synthesized in the laboratory, the result is an equal (racemic)
mixture of both L- and D-forms, like a pile of left- and right-
handed gloves. Yet living things include only "right-handed"
(D-) sugars. How did they come to use only one form?
Scientists have conducted experiments to determine how these
preferences might arise by material mechanisms. So far, they
have found no mechanism that will produce only the correct
three-dimensional structure. When material mechanisms
(such as Hazen's calcite crystals) are applied to such mixtures
to segregate L- from D-forms, the result is at best a slight disproportion of one form over the other. In this and other ways,
life shows characteristics that are alien to anything produced
under ordinary material conditions.

Oparin's assumption that biologically relevant molecules such as amino acids and sugars could assume the right uniform orientation by purely material means (Assumption #4 in chapter 4) therefore also appears to be incorrect because no such naturally occurring mechanism appears to exist.

The Synthesis of Polymers

Once sugars and amino acids of the right chirality (or handed-ness) are concentrated in the right place, the real work of constructing life begins: L-amino acids need to be combined in the right sequence to form proteins, and nucleotide bases attached to a (D-) sugar-phosphate backbone need to be combined in the right sequence to form DNA and RNA. Proteins, DNA, and RNA are long polymer chains whose biological function depends on the precise sequence of monomers that make up the chains (an average protein is several hundred amino acids in length; DNA can be millions of nucleotide bases in length).

Synthesizing polymers with the right sequences requires two steps: (1) getting adjacent monomers to join together properly and (2) getting properly joined monomers to assume a functionally meaningful order. On the first step, because of the ways that amino-acid and nucleotide monomers react chemically, they can join in numerous ways. Amino acids, for instance, can react with each other to form a variety of chemical bonds. Yet only one, the *peptide bond,* occurs in functional proteins. Amino acids joined by peptide bonds do not necessarily form a functional protein, but all chains that are functional feature peptide bonds. The chain must also be linear. Branching chains, in which amino acid residues joined by peptide bonds form side chains can also occur, but they will not work. Amino acid chains in proteins must be linear.

To see how this works, consider an analogy based on the following sequence of letters:

ANERUT

Ordinarily, when we arrange letters in a sequence, we orient them vertically from the top down. This orientation corresponds, in our analogy, to amino acids joined exclusively by peptide bonds. As we arrange letters in sequence, however, we could also rotate them. Thus, we might not only rearrange the order of the previous letters but also rotate them:

Such rotations correspond to nonpeptide bonds in a linear sequence of amino acids.[1]

Rotating the letters in odd ways like this disrupts any meaning that the sequence might have if the letters were given their proper vertical orientation (for instance, is the first letter a "Z" or an "N"?). Likewise, an amino acid sequence that would form a functional protein if all the bonds were peptide bonds would have its function disrupted by nonpeptide bonds. Thus, for the letters A, N, E, R, U, and T to form a meaningful word, the letters must be both properly oriented and correctly ordered:

NATURE

Likewise, to form a functional protein, amino acids must be both properly linked (by peptide bonds) and correctly ordered (to ensure that the resulting amino-acid sequence folds).

Nucleotide sequences face exactly the same constraints: in the polymerization of polynucleotides, which is the primary step in the formation of DNA and RNA, 3–5 phosphodiester linkages are the only ones that cellular life allows even though

2–5 linkages dominate in the absence of cellular life (and thus in prebiotic environments). It follows that without some catalyst that promotes peptide bonds (for amino-acid sequences) or 3–5 phosphodiester linkages (for nucleotide sequences), there can be no materialist route to proteins, DNA, and RNA. But the only catalysts we know capable of handling this task are enzymes and other protein-based products (e.g., the ribosome). These catalysts in turn require the entire DNA-RNA-protein machinery. This machinery, however, is precisely what origin-of-life research may not presuppose but rather must explain.

To describe the requirements for obtaining functional amino-acid or nucleotide sequences does not explain how such sequences originated. According to the Oparin hypothesis, such sequences assembled themselves gradually through an undefined process of self-organization. But this answer is unconvincing. The probability of obtaining one specific protein in an undirected search of amino-acid sequences is practically zero. Consider a small protein consisting of 100 amino-acid subunits. How many different sequences of the basic 20 amino acids are possible in a chain of 100 subunits? The answer is 20100, or approximately 10130 (1 followed by 130 zeros). The improbability of finding such a small protein by a blind or random search of protein sequence space is therefore 1 in 10130. The number 10130 is so enormous that even with billions and billions of years there have not been enough opportunities for the known physical universe to sort through every combination of amino acids to find the specific combination of amino acids for even one such protein.[2]

These numbers can be whittled down, but not drastically. A chain of amino acids tolerates substantial variation at sites along the chain without disrupting the protein's function. MIT biochemist Robert Sauer has applied a technique known as *cassette mutagenesis* to several proteins to determine how much variation among amino acids can be tolerated at any given site. His results show that even with this variation taken into account, the probability of forming a 100-subunit functional protein is only 1 in 1065—this is still a vanishingly small probability (there are estimated to be 1065 atoms in our galaxy).[3]

Building on Sauer's work, Douglas Axe developed a refined mutagenesis technique while at Cambridge University to measure the sequence specificity of enzymes such as barnase. Axe's research indicates that previous mutagenesis experiments actually underestimated the functional sensitivity of proteins to amino-acid sequence change because they presupposed (incorrectly) the context independence of function-preserving changes at individual amino-acid sites.[4] The probability of achieving a functional 100-subunit sequence of amino acids by undirected search is therefore considerably smaller than the 1 in 1065 improbability calculated by Sauer.[5]

These estimates of improbability are extremely conservative, giving prebiotic chemistry every unlikely advantage to succeed. Modern organisms use just 20 basic amino acids. But the evidence we have for a "prebiotic soup" (e.g., from simulation experiments and from studying amino-acid-rich meteorites) indicates that it would have contained many more than these 20. The Murchison meteorite, for instance, has been found to contain over 70 amino acids.[6] Any realistic origin-of-life scenario attempting to account for the polymerization of amino acids is therefore likely to contain at least 50 amino acids, each with right- and left-handed forms. In a sequence of amino acids, that's 100 amino-acid possibilities at each location. Thus, for an arrangement of 100 amino acids, there are 100100 = 10200 possibilities. And that assumes only linear arrangements and only peptide bonds. Even this estimate, however, is conservative. A more realistic estimate would include not only these additional amino acids but also reactions with countless competing compounds (sugars, aldehydes, etc.). The number of possibilities, and the corresponding improbabilities, quickly become staggering.

How do origin-of-life researchers handle such improbabilities? Do they treat them as possible evidence for design? No. Typically they respond by charging that design theorists have concocted a "tornado in a junkyard." That is, design theorists merely calculate the probability of functional protein types forming by purely random assembly. Instead, origin-of-life researchers urge, evolutionary mechanisms exist that substantially raise the probability of the formation of these systems. But what is the evidence for the existence and effectiveness of such mechanisms? It is nil.

Origin-of-life researchers have proposed speculative scenarios in which mechanisms of chemical evolution supposedly create biologically significant polymers (see chapters 11, 15, and 19). But these scenarios don't even begin to address the true complexities of life. Moreover, origin-of-life researchers have failed to test and confirm these scenarios experimentally. Oparin boldly assumed the chemical pathways he needed to account for the origin of life, but then left the crucial demonstration of those pathways to others. Not only are we still waiting for such a demonstration, but, as we have just seen, the chemical pathways we do know conspire against the formation of functional polymers such as proteins and DNA.

Oparin's assumption that protein and nucleotide polymers could gradually form by purely material processes acting on monomers (Assumption #5 in chapter 4) therefore also appears to be incorrect.

12

The Humpty-Dumpty Problem

Take a clean, sterile Eppendorf tube (a small test tube with a tapered bottom that is often used by biochemists). Place into it a small amount of sterile salt solution, known as a buffer, at just the right temperature and pH. Place into the buffer a living cell, but puncture it with a sterile needle so that its contents leak out into the solution. (The amount of buffer can be kept as small as possible to avoid diluting the cell's contents too much.) You now have all the molecules needed to make a living cell—not just the building blocks, but all the fully assembled macromolecules such as DNA and proteins. Moreover, you have them in just the right proportions, under just the right conditions, and without any interfering substances. Yet, despite all such efforts to facilitate the transition from nonlife to life, the molecules will not form into a living cell—no matter how long you wait or what you do to them, given present knowledge and technology. What's more, if our technology ever gets to the place where it can reassemble punctured cells, that will be evidence not for the power of chemical self-assembly but for design.

We've just sketched what may be called the Humpty-Dumpty experiment. Origin-of-life researchers, playing the role of "all the king's horses and all the king's men," have been spectacularly unsuccessful in putting Humpty-Dumpty back together again. But if reassembly has proven difficult, think of how much more difficult it is to assemble Humpty-Dumpty from scratch, as the Oparin hypothesis assumes and origin-of-life researchers hope to demonstrate. To put the problem in perspective, focus for the moment simply on the enzymes required for protein synthesis. Enzymes are proteins that specialize in carrying out chemical reactions quickly and efficiently. The enzymes in a cell work together like the parts of a finely crafted machine to perform the cell's metabolic functions. A typical enzyme consists of several hundred amino acids.

To build even a single protein in a cell, about sixty specific proteins acting as enzymes are needed. These include helicases (which unwind double-stranded DNA into single strands), RNA polymerases (which polymerize RNA with respect to a DNA template), aminoacyl-tRNA synthetases (which link amino acids to transfer RNA), and ribonucleases (which control the amount of unprotected RNA in the cell). These sixty enzymes need to act in concert with the cell's genetic machinery—its DNA. If even one of these enzymatic proteins necessary for protein synthesis is missing, the cell will not be able to form proteins. The enzymes involved in protein synthesis are therefore indispensable for life and, considered jointly, form an irreducibly complex system (see chapter 22).

The precise division of labor in a cell's protein-making machinery underscores the enormity of the problem of evolving proteins under prebiotic conditions. Protein synthesis, which is essential for life as we know it, presupposes these highly specific enzymes. Moreover, not only must those sixty enzymes all exist at the same time; they must also occur together in the same tiny region of the cell. In other words, they must be coordinated and targeted to the right location in the cell. But such coordination itself presupposes a vastly complex communication, transportation, and control system (and that system is itself enzymatically, and thus protein, driven).

Consequently, getting sixty specific enzymes to congregate and act in precisely choreographed steps in the correct sequence and in the exact location still doesn't master the full enormity of the problem. The cellular environment for these proteins must itself be conducive to their coordinated function. And this presupposes that all basic cellular functions be in place from the start. These include: the storage, retrieval, and processing of the information in DNA; the manufacturing of proteins from RNA templates by ribosomes (ribosomes are themselves immensely complex, consisting of at least fifty protein and RNA subunits); the metabolic functions that extract energy from nutrients; and the very capacity to reproduce such vastly complex systems.

Oparin's assumption that the biomacromolecules could gradually be coordinated and integrated into a primitive life form by purely material processes (Assumption #6 in chapter 4) therefore also appears to be incorrect.

13

Harnessing the Sun

Oparin's final assumption is that some cells eventually developed the ability to capture sunlight and transform it into usable energy by means of photosynthesis. Photosynthesis is the process by which plants and certain types of single-celled life forms convert the energy from light, carbon dioxide, and water into reduced carbon and oxygen. It is almost the exact reverse of oxidation, and is critical for producing the fuel needed for complex life forms such as animals and plants.[1]

Cyanobacteria (blue-green algae) are believed to have evolved the complex process of photosynthesis in the Archean period (3.8–2.5 billion years ago), but no adequate explanation exists for how that evolution occurred, only "competing hypotheses."[2] Presently known material mechanisms fail to account for how so complex a process as photosynthesis developed among these early life forms. In particular, no scientist has proposed any detailed step-by-step pathways for the evolution of photosynthesis.

To be sure, biologists have speculated how photosynthesis might have evolved.[3] But such speculation remains for now untestable. Moreover, positive reasons now exist for thinking that photosynthesis was designed. Indeed, the exquisite fine-tuning of photosynthetic complexes to capture solar light efficiently by exploiting quantum-mechanical effects and then by transferring that energy to reaction centers in the cell where it can be stored and later utilized suggests a level of design that far exceeds anything produced thus far from the field of photochemical engineering.[4]

Oparin's assumption that primitive cells could have developed photosynthesis by purely material processes (Assumption #7 in chapter 4) therefore appears to be incorrect as well.

14

The Proteinoid World

Most simulation experiments do not realistically model the conditions that our best evidence suggests existed on the early Earth. In the most likely scenario, the primitive atmosphere contained free, destructive oxygen; reactions promoting life tended easily to reverse themselves; interfering cross-reactions were common; and right-handed amino acids and left-handed sugars appeared in the same proportion as their biologically useful counterparts. So there is no reason to think that conditions on the early Earth favored the emergence of life's basic building blocks by purely material forces.

The most probable scenario of early-Earth history does not build up biological complexity but rather breaks it down.[1] Moreover, there is neither a solid theory nor a promising experimental basis for the view that biomacromolecules could have emerged by purely material forces, much less that an interdependent and coordinated set of them could gradually organize themselves into functionally integrated units with the complexity of actual cells. Oparin's entire framework for the origin of life is therefore thrown into question.

In critiquing the Oparin hypothesis, we have thus far focused on the obstacles to making and arranging the building-blocks required to form complex biomolecular structures. Though these obstacles are daunting and, in our view, sufficient to invalidate the whole materialist origin-of-life enterprise, origin-of-life researchers have not given up. Instead, they have come up with proposal after proposal for fleshing out Oparin's hypothesis in the hopes of overcoming these obstacles. In this and subsequent chapters we turn to some of these proposals in detail. One of the earliest was the late Sidney Fox's proteinoid microspheres. Fox wrestled with how purely material forces could construct complex, information-rich biological structures from elementary constituents.

How, for instance, did amino acids come together to form the first proteins? Amino acids, left to themselves, tend to clump together, but not with the right bonds and not in the right order to form proteins. For that, they need enzymes that act as specific catalysts and DNA to serve as a sequence template. But these enzymes themselves are very special proteins that require DNA to code for their catalytic ability. Moreover, DNA itself requires enzymes for its production. How, then, did the first amino acids form enzymes when there were neither existing enzymes to serve as catalysts in linking the protein structure together nor DNA to code for the structure?

This is a classic case of *Which came first, the chicken or the egg?* To overcome this problem, scientists have attempted to link amino acids together into protein-like molecules without using either existing enzymes or DNA. Fox, while at the University of Miami, performed an experiment to that end, heating dry mixtures of amino acids at 160–180° C for several hours in a nitrogen atmosphere. He found that these amino acids joined together. Nevertheless, they did not form linear protein polymers. His polymers included nonpeptide bonds and were branched. To call them polypeptides would therefore have been incorrect. Fox dubbed them "proteinoids" instead. It would have been more accurate to call them "branched amino-acid polymers."

Fox's proteinoids faintly exhibited some of the properties of true proteins. For instance, the proteinoids catalyzed certain chemical reactions, so that they speeded up slightly. Fox contended that these proteinoids represented the beginnings of enzymatic activity on the primitive Earth. Moreover, if proteinoids are dissolved in boiling water and the solution is cooled, the proteinoid molecules will clump together to form uniform microscopic spheres about the size of bacterial cells.

Fox contended that these little spheres represent the first step toward cellular life on Earth.

The process of forming proteinoid microspheres from amino acids is thought to have taken place near volcanoes (this assumes, of course, that amino acids could have accumulated in sufficient numbers on the primitive Earth—this assumption is itself problematic, as we saw in chapters 7 through 10). As Fox conceived it, amino acids formed when gases of the primitive atmosphere came in contact with molten rock from a volcano (at a temperature of about 1200° C). The amino acids then accumulated some distance from the center of the active volcano where surface temperatures ranged between 160 and 180° C. At such temperatures, the dry amino acids condensed into proteinoids. These were then protected from the destructive effects of heat by the cooling action of rainfall.

Once suspended in pools of water, the proteinoids would form microspheres. These in turn might acquire the ability to compete with each other for environmental resources (how they might do this without first acquiring the ability to metabolize nutrients and reproduce is unclear—the proteinoid microspheres that emerged from Fox's experiment were incapable of either). The most successful competitors might then survive, become dominant, and develop life-like qualities. The process could take millions of years, though Fox suggested it could take but hours to get to the protocell stage. Eventually, the first self-reproducing cells capable of fermentation would form, and life itself would gain its first foothold.

Fox's experiment seems to reach the threshold of life by transforming amino-acid building blocks, which can be produced in Miller-Urey type simulation experiments. Taken together, the two experiments might seem to confirm Oparin's hypothesis: Miller-Urey type experiments supply the building blocks, and Fox's proteinoid microspheres assemble them into life-like wholes. Closer examination, however, reveals that Fox's experiment falls far short of providing a credible scenario for the origin of life.

There are several problems with Fox's experiment. First, Fox used mixtures containing only protein-forming L-amino acids. Where on the primitive Earth, or elsewhere, could such a mixture have occurred? As we have seen, naturally occurring amino acids always include roughly the same number of L- and D-amino acids. Also, even with a mixture including only L-amino acids, interfering cross-reactions would have tied up the amino acids in nonbiological compounds, thus blocking the formation of proteinoids. Sugars, for instance, react with amino acids to form the nonbiological compound known as melanoidin. Because of such cross-reactions, which Oparin's model neglected, it is very unlikely that proteinoids could have formed under natural conditions on the primitive Earth. As a consequence, many scientists regard Fox's use of selected and purified amino-acid mixtures as highly unrealistic.[2]

Second, the proposed sequence of events that supposedly occurred near volcanoes is dubious. The required combination of high and low temperatures, with rainstorms occurring at just the right time and place, seems unrealistically "choreographed." Moreover, the heat needed to form proteinoids would also have threatened to destroy them. And if the heat did not destroy them, the proteinoids would likely have disintegrated on their own before they could play a role in the formation of life.

Third, unlike Fox's proteinoid microspheres, actual cells are surrounded by a complex cell membrane. Cell membranes are made of specialized fatty acids and phosphate groups that form a phospholipid bilayer, together with many specialized protein molecules. Unless the protein molecules are highly specific, they will not function properly in forming and maintaining the cell membrane. By contrast, proteinoid microspheres have an outside boundary that is much thicker than the membrane of actual cells. Moreover, the microspheres are composed entirely of proteinoids, very simple molecules compared to true proteins.

Last, and most significantly, even if the preceding problems could be resolved, so that proteinoid microspheres could readily be produced under realistic prebiotic conditions, massive differences exist between them and the very simplest living cells.[3] Unlike proteinoids, the simplest cell requires the highly organized arrangement of hundreds of precisely defined information-rich macromolecules (e.g., proteins and DNA) for its chemical machinery to accomplish all the specialized functions a cell needs to live and prosper. Fox's proteinoid microspheres provide no mechanism for the formation of these biomacromolecules and therefore leave their origin completely unexplained.

To recap, we have considered four problems with Fox's experiment: (1) his use of selected and purified amino-acid mixtures (only protein-forming L-amino acids); (2) his implausible scenario for the formation of proteinoid microspheres near volcanoes; (3) his identification of proteinoid microspheres with membrane-enclosed cells even though the two are vastly different; and finally (4) his unfounded claim that the minimal catalytic activity of proteinoids might somehow lead to the biomacromolecules relevant for life and that these in turn might gradually organize themselves into actual cells. Each of these problems alone shows that Fox's experiment cannot adequately account for the origin of life. Together they show that Fox's proteinoid world is not a plausible route to life's origin.

The RNA World

The origin-of-life scenario that has attracted the most attention since the 1990s is the "RNA world" (also known as "ribozyme engineering"). The RNA world addresses a key weakness of hypotheses we have considered till now. Primitive atmosphere simulation experiments such as those by Miller and Urey and proteinoid experiments such as those by Sidney Fox, even if they could explain the formation of life's elementary constituents (e.g., amino acids), nonetheless fail to explain how proteins could have formed on the early Earth.

Before the RNA world was proposed, biologists tended to elevate proteins as the master molecule from which all other biomacromolecules arose. Placing proteins in this privileged position, however, leads to a vicious circle: proteins presuppose DNA and DNA presupposes proteins. As a consequence, origin-of-life researchers have considered the possibility that proteins were not the first molecular building blocks of life. DNA is not a good candidate to be the first, however, because it needs a whole array of complex proteins to make copies of itself.[1] Consequently, DNA could not have originated before proteins and could not have been the first step in the origin of life.

But what about RNA? RNA is a close chemical relative of DNA that is used by all living cells to make proteins. In the 1980s molecular biologists Thomas Cech and Sidney Altman showed that RNA can sometimes behave like an enzyme—that is, like a protein that catalyzes chemical reactions. Keying off this discovery, molecular biologist Walter Gilbert suggested that RNA might be able to synthesize itself in the absence of proteins, and thus might have originated on the early Earth before either proteins or DNA. This "RNA world" might then have been the molecular cradle from which living cells emerged. RNAs with catalytic properties became known as "ribozymes," and the search for such RNAs became known as "ribozyme engineering."

But this scenario is deeply problematic as well. No one has been able to demonstrate convincingly how RNA could have formed before living cells were around to make it. According to Scripps Research Institute biochemist Gerald Joyce, RNA is not a plausible candidate for the first building block of life "because it is unlikely to have been produced in significant quantities on the primitive Earth."[2] RNA consists of a nucleotide base (either adenine, cytosine, guanine, or uracil), the five-carbon sugar ribose, and a phosphate group. The material processes that favor the formation of these nucleotide bases work against the formation of the corresponding sugar and phosphate group, and vice versa. Indeed, the experimental synthesis of RNA has to date occurred only under the most unrealistic prebiotic conditions. NYU biochemist Robert Shapiro describes one such experiment:

> [O]ne example of prebiotic synthesis [was] published in 1995 by *Nature* and featured in the *New York Times*. The RNA base cytosine was prepared in high yield by heating two purified chemicals in a sealed glass tube at 100 degrees Celsius for about a day. One of the reagents, cyanoacetaldehyde, is a reactive substance capable of combining with a number of common chemicals that may have been present on the early Earth. These competitors were excluded. An extremely high concentration was needed to coax the other participant, urea, to react at a sufficient rate for the reaction to succeed. The product, cytosine, can self-destruct by simple reaction with water. . . . Our own cells deal with it by maintaining a suite of enzymes that specialize in DNA repair. The exceptionally high urea concentration was rationalized in the *Nature* paper by invoking a vision of drying lagoons on the early Earth. In

a published rebuttal, I calculated that a large lagoon would have to be evaporated to the size of a puddle, without loss of its contents, to achieve that concentration. No such feature exists on Earth today.[3]

We hasten to add there's no evidence that any such feature existed in the distant past either. But even if RNA could be produced under realistic prebiotic conditions, it would not survive long under the conditions thought to have existed on the early Earth—it is simply too unstable. Joyce therefore concludes: "The most reasonable interpretation is that life did not start with RNA."[4] Although he still thinks that an RNA world preceded the protein-DNA world, he believes that some kind of non-RNA life must have preceded RNA-life (that then evolved into DNA-life). According to Joyce, "You have to build straw man upon straw man to get to the point where RNA is a viable first biomolecule."[5]

Even if RNA were a viable first biomolecule, the RNA world has yet to suggest a self-consistent, self-contained materialist account of life's origin. Consider a standard experiment from the ribozyme engineering literature: SELEX. The acronym SELEX stands for "*S*ystematic *E*volution of *L*igands by *EX*ponential enrichment." In 1990 the laboratories of J. W. Szostak (Boston), L. Gold (Boulder), and G. F. Joyce (La Jolla) independently developed this technique, which permits the simultaneous screening of more than a thousand trillion (i.e., 1015) polynucleotides for different functionalities (polynucleotides are sequences of DNA or RNA).[6]

A typical SELEX experiment starts with a random pool of RNAs that cannot do much of anything and ends with RNAs that can perform a particular function, such as catalyzing a specific reaction or binding to a specific molecule. Consequently, there appears to be a net increase in biologically useful information over the course of the experiment. Moreover, the molecules one gets at the end of the experiment do not match any blueprint identifiable in advance. Thus, the experimenter cannot predict the precise molecular structures that emerge.

An extensive effort usually follows a SELEX experiment to characterize the evolved RNA. The RNA must be sequenced, and in some cases it is crystallized for the structure to be solved. Only then does the scientist know what was created and how it performs its function.

SELEX experiments mimic Darwinian evolution in the sense that RNAs that approximate some function get selected and then preferentially duplicated. Do SELEX experiments therefore demonstrate the power of purely material forces to evolve biologically significant RNA structures under realistic prebiotic conditions? Not at all. Intelligent intervention by the experimenter is indispensable. In SELEX experiments large pools of randomized RNA molecules are formed by intelligent synthesis and not by chance—there is no natural route to RNA (in fact, the chemical processes in nature that facilitate the formation of nucleotide bases undercut the formation of RNA's sugar-phosphate backbone and vice versa). The artificially synthesized molecules are then sifted chemically by various means to select for catalytic function. What's more, the catalytic function is specified by the investigator. Those molecules showing some activity are isolated and become templates for the next round of selection. And so on, round after round.

At every step in SELEX and ribozyme (catalytic RNA) engineering experiments, the investigator is carefully arranging the outcome, even if he or she does not know the specific sequence that will emerge. It is simply irrelevant that the investigator learns the identity or structure of the evolved ribozyme only after the experiment is over. The investigator first had to specify a precise catalytic function. Next, the investigator had to specify a fitness measure gauging degree of catalytic function for a given polynucleotide. And finally, the investigator had to run an experiment to optimize the fitness measure. Only then does the investigator obtain a polynucleotide exhibiting the catalytic function of interest. In all such experiments the investigator is inserting crucial information at every step. Ribozyme engineering is engineering. Indeed, there is no evidence that material processes as found in nature can do their own ribozyme engineering without the aid of human intelligence.

In short, the RNA world therefore offers no explanation for how such a "viable first biomolecule" might have arisen. But suppose some as-yet-unknown materialist scenario could have produced such a molecule.[7] What would this molecule look like? And what would it be capable of evolving into? Would it be a single self-replicating RNA? If so, what happens next? Would more and more RNAs accumulate to form a coordinated system of self-replicating RNAs? And how would such a system evolve into the DNA-protein machinery that is standard issue for life as we know it? The RNA world leaves all such questions unanswered. Indeed, origin-of-life researchers have no clue about how to answer them. The RNA world—like the protein-first scenario in the Miller-Urey and Fox experiments—therefore presupposes a still more fundamental origin-of-life scenario.

Self-Organizing Worlds

Given the daunting challenges facing a materialist explanation of life's origin, one might think that origin-of-life researchers would be pessimistic about their prospects for resolving this problem. But one would be mistaken. Origin-of-life researchers tend to be optimistic about their prospects for ultimate success, at times even giving the impression that, except for a few minor details, the problem of life's origin has essentially been solved already. As an example of such optimism, consider the following remark by mathematician Ian Stewart:

> The origin of life no longer appears to be a particularly difficult problem. We know that—at least on this planet—the key ingredient is DNA. Life's basis is molecular. What we need is an understanding of complex molecules: how they might have arisen in the first place, and how they contribute to the rich tapestry of living forms and behavior. It turns out that the main scientific issue is *not* the absence of any plausible explanation for the origin of life—which used to be the case—but an embarrassment of riches. There are many plausible explanations; the difficulty is to choose among them. That surfeit causes problems for the question "How *did* life begin on Earth?" but not for the more basic issue, which is "*Can* life emerge from nonliving processes?"[1]

Stewart's assessment of the origin-of-life problem is too cavalier. Indeed, the embarrassment of riches that Stewart cites should give us pause. It is usually hard enough to come up with even one good theory to account for a phenomenon. For instance, prior to Newton's theory of universal gravitation, there was no unified theory to account for the motion of stars, moons, and planets. An embarrassment of riches points not to the solution of a problem but to vain gestures at a solution.[2] Indeed, the very claim that "there are many plausible explanations" suggests that none is compelling. If any one of them were compelling, we could expect it to become a consensus among scientists (to say nothing of it being considered the "true" or "correct" explanation). Instead, we find a vast proliferation of "plausible explanations" none of which is universally acclaimed and each of which has fatal flaws.

Whenever origin-of-life researchers accept plausibility rather than evidence as their standard for scientific truth, they in effect give up the search for what really happened or for what with reasonable probability could have happened. *Science fiction seeks plausibility; science seeks probability.* Plausibility, as Stewart and many origin-of-life researchers understand the term, implies no effort to estimate probability. Instead, they settle for what they can *imagine* was possible or could have happened. In this way, they substitute opinion and prejudice for experiments and data. Where is the science in all this? How is the plausibility that Stewart ascribes to the various materialist accounts of life's origin anything other than an article of faith?

Even so, Stewart raises two questions that deserve closer scrutiny: (1) How did life begin on Earth? and (2) Can life emerge from nonliving processes? Stewart regards the answer to the first question as a work in progress and the answer to the second as an unequivocal yes. But what is this second question really asking? If it is merely asking whether nonliving matter can be organized into living matter, the answer is clearly yes. When broken down far enough, all matter consists of particles that can exist on their own apart from life. Indeed, cosmologists have determined that all matter on Earth once belonged to stars, and it was not living then.

In saying that life can emerge from nonliving processes, Stewart is saying that nonliving matter, without any outside assistance, has the ability to organize itself into living matter.

Indeed, his reference to life's "emergence" denotes, within the origin-of-life community, the power of nonliving matter to organize itself into living matter not as some wildly improbable event but as an inherent feature of nonliving matter. Thus, origin-of-life researcher Harold Morowitz, in his book *The Emergence of Everything,* writes,

> The view of the emergence of biochemistry that we have been discussing represents a paradigm shift from what the reader may have encountered in biology courses where it was assumed that random products of free-radical reactions lead to monomers, then to polymers, then to cells [cf. the RNA world discussed in chapter 15]. In the view elaborated here, selection rules lead to a core metabolism that then produces an ordered hierarchy of emergent structures and functions. These become increasingly complex, leading to the sophisticated chemistry of the universal ancestor. This is a very different view than may have been taught in standard introductory courses, but I believe that it is a *much more probable scenario.*[3]

Morowitz here contrasts his own "metabolism-first" approach to life's origin with the more conventional "genetics-first" approach that regards the key to life's origin as the synthesis of biologically functional polymers (as in the RNA world). Although the genetics-first approach has to date received the most attention from origin-of-life researchers, it suggests that at key points in life's origin, as monomers were being configured into polymers, vastly improbable events had to occur.[4] This is because functional polymers tend to be rare among all possible polymers; moreover, the physical properties of monomers that are relevant to their linear arrangement exhibit no preference for one sequence over another.

For instance, with RNA, the sugar-phosphate backbone onto which the nucleotide bases get attached is completely indifferent to how those bases are ordered (much as a refrigerator does not care about the order in which children place magnetic letters on the door). The ability of polymers to carry information results from the high degree of freedom that the laws of physics and chemistry permit *to the monomers in the order in which they may combine.* These laws prefer no sequence over any other. Yet precisely because of this freedom, there is no way for monomers to organize themselves spontaneously into biologically functional polymers except as vastly improbable events (unless they are extremely short; but in that case what these polymers gain in probability is offset by their loss of biological function).

The problem with vast improbabilities is this: If the origin of life is a vastly improbable event, then it is a freak—an exceedingly lucky event that happened but need not have happened and is far more likely not to have happened. But to say that life's origin is lucky does not constitute a scientific theory. A scientific theory of life's origin is possible only if life originated through a sequence of steps where each step is probable and the sequence as a whole is also probable. In that case, nonliving matter has an inherent propensity to organize itself into life. In other words, it constitutes a self-organizing world. Morowitz's metabolism-first approach, in which a core metabolism facilitates the emergence of increasingly complex structures and functions, is therefore an instance of a self-organizing world.

- Besides Morowitz's self-organizing metabolism-first world, origin-of-life researchers have proposed many other self-organizing worlds. Here is a representative sample:[5]

- Stuart Kauffman, seeing self-catalyzing collections of chemicals as the key to life's origin, has proposed that "life emerged as a phase transition to collective autocatalysis once a chemical minestrone, held in a localized region able to sustain adequately high concentrations, became thick enough with molecular diversity."[6]

- Christian de Duve, in assuming that chemistry is prior to information, regards the origin of life as a cosmic imperative (i.e., as an outcome destined by the very nature of nature) and has proposed thioesters (certain sulfur compounds) as the basis for metabolism and thus as the key to life's origin.[7]

- Günter Wächtershäuser, looking to bacteria that consume iron and sulfur, has proposed the mineral pyrite (which is rich in iron and sulfur and is commonly known as "fool's gold") as a primary catalyst for the origin of life.[8]

- Michael Russell has proposed that life originated as convection currents produced by undersea volcanic vents set up a temperature differential between competing fluid flows at which colloidal membranes made of iron sulfide could form.[9]

- David Deamer has proposed that the self-organization of lipids into bilayered vesicles, which resemble cell membranes, was a crucial factor in the origin of life.[10]

- Simon Nicholas Platts has proposed that life originated through the activity of polycyclic aromatic hydrocarbons (abbreviated PAHs) which, according to his model, supply "proto-informational templating materials" and thus serve as progenitors for the RNA world.[11]

- Graham Cairns-Smith has proposed a clay-template theory for the origin of life in which self-replicating clays form templates for subsequent carbon-based life.[12]

All such proposals are highly speculative, address only one or a few of the most elementary aspects of the origin of life, are thin on detail even in those aspects of the origin of life that they do address, have little if any experimental support, and require massive intervention from investigators to achieve any interesting results (thereby failing to reflect realistic prebiotic conditions).[13]

Harold Morowitz, for instance, has for years been trying to get prebiotic chemicals to organize themselves into the citric acid cycle.[14] This cycle, as with other metabolic cycles, occurs

in all cells and exhibits a form of self-replication (in metabolic cycles, one chemical catalyzes another, which catalyzes
another, and so on, until the process circles back to the first
chemical, producing more of it than was originally there and
thus replicating the cycle). For Morowitz, the emergence of
the citric acid cycle is a key step in the origin of life. Yet, to
this day, he cannot obtain the citric acid cycle except with the
help of enzymes,[15] which presuppose life as we know it and
therefore properly speaking do not belong in the origin-of-life
researcher's tool-kit.

Or take David Deamer's "lipid world." Deamer has demonstrated experimentally that carbon-based molecules taken
from the Murchison meteorite (which presumably represents
a pristine prebiotic environment) spontaneously organized
themselves into small spheres the size of microbes. Moreover,
the spheres consisted of lipid bilayers as found in cell membranes. So here we see another key step in the origin of life
attributable to purely material factors. Or do we? "Biologists
have been quick to point out," notes Robert Hazen, that "the
vesicles produced in Deamer's work are a *far cry* from actual
cell membranes, which feature a mind-boggling array of protein receptors that regulate the flow of molecules and chemical
energy into and out of the cell."[16] Many other pieces of life's
puzzle would need to fall into place before we could say with
any assurance that Deamer's lipid world plays a pivotal role in
the origin of life.

Even if we bend over backwards to be as charitable as
possible to these self-organizing worlds, they remain deeply
problematic. Suppose they were to succeed on their own terms.
What exactly would that show? What exactly would have
organized itself? And how would such products of self-organization be significant to the larger story of life's origin? The
origin-of-life community has already conceded that catalytic
RNAs cannot organize themselves from their elementary constituents. Ditto for DNA and proteins. At best, self-organizing
worlds therefore yield primitive replicators that look nothing
like DNA, RNA, or protein. But in that case, why should we

think that such replicators have any relevance to the origin of cellular life? Metabolism-first worlds, for instance, never describe the route by which primitive metabolisms achieve the ability to arrange nucleotides into functional sequences (as required by the RNA world).

A spectacular success for Wächtershäuser's iron-sulfur world model would be to demonstrate how a metabolic cycle could organize itself by employing only authentic prebiotic materials under authentic prebiotic conditions (and hence without excessive investigator interference). Although Wächtershäuser and his colleagues have accomplished nothing like this,[17] imagine that they had. What form would such "primitive metabolic life" take? It would be a coating on a mineral—perhaps a glaze or patina. Yet why should we think that such a coating might play a pivotal role in life's origin? We have no compelling reason to draw such a conclusion short of a detailed evolutionary path from such a coating to cellular life. But no such paths are known (not even paths terminating in the much simpler RNA world are known).

In thinking that Wächtershäuser's iron-sulfur world played a pivotal role in bringing about the actual world of cellular life, we might just as well think that waterwheel technology played a pivotal role in bringing about supercomputer technology. The supercomputer owes nothing to the waterwheel. This lack of connection is also evident with Wächtershäuser's model as well as with the others considered in this chapter. To think that cellular life "emerged" from any of these "worlds"—in the absence of hard evidence—is wishful thinking.

Like the alchemists of old, who never explained exactly how gold could be produced from lead, origin-of-life researchers never tell us exactly how "an ordered hierarchy [of] structures and functions"[18] could emerge from the chemical processes that might reasonably have existed on the prebiotic Earth. They claim that matter has the ability to transform itself in remarkable ways, yet without identifying the precise causal pathways by which it could. "X emerges" is an incomplete sentence. It needs to read "X emerges from Y." And even in this form it remains incomplete until one precisely specifies Y and provides a detailed account of how Y could in fact produce X.[19] Moreover, such an account needs to be backed up by evidence. Otherwise, to invoke emergence as an explanation constitutes a leap of faith.

A complete set of the building materials for a house does not account for a house—additionally what is needed is an architectural plan (drawn up by an architect) as well as assembly instructions (executed by a contractor) to implement the plan. Likewise, with the origin of life, it does no good simply to describe a plausible set of chemical precursors to life. A detailed account of how purely material forces could, under plausible prebiotic conditions, organize those precursors into a living organism needs to be specified as well. Such accounts don't exist. Notwithstanding, origin-of-life researchers continue to regard self-organizing worlds as deeply relevant to the origin of life. To understand why, we need to turn to our next topic: molecular Darwinism.

Molecular Darwinism

Origin-of-life researchers have one remaining strategy for explaining how life originated by purely material forces, namely, to apply the Darwinian mechanism not to organisms but to molecules. "Molecular Darwinism,"[1] as we may call it, is thereby supposed to explain the origin of life. What if researchers could identify a simple self-replicating molecule or set of molecules? Suppose such a molecular system is so simple that it could arise from one of the self-organizing worlds described in the last chapter. Might it not be possible for the Darwinian mechanism to do the rest? Why shouldn't natural selection and random variation act at the molecular level, bootstrapping that initial replicator all the way up to a full-blown cell?

There is good reason to be skeptical about the effectiveness of natural selection before the advent of cellular life. Origin-of-life researchers do not need to explain the origin of simple molecular replicators, but rather the origin of a self-replicating system of macromolecules that can perform the very specific functions universally associated with life as we know it (material transport, metabolism, energy conversion, signal transduction, information processing, sequestration from the environment, etc.). Origin-of-life research is constantly lowering the bar for what may count as first life. Not replication per se, but self-replicating multimolecular systems of the kind found in actual cells are what need to be explained.[2]

For instance, we have long known that a crystal seed placed in a supersaturated solution will cause crystal growth. In this way, the crystal seed may be said to self-replicate. Even so, self-replication here bears little resemblance to the self-replication of functionally integrated multimolecular systems found in cellular life. It is such systems that molecular Darwinism needs to explain. As a consequence, even if one were to concede that a relatively simple self-replicating molecule or set of molecules could arise by purely material processes (that is, without extensive investigator interference), the Darwinian mechanism of natural selection and random variation operating at the molecular level would still need to explain

- how these molecules could become enclosed in a channeled and gated membrane;

- how they could generate an ever-increasing assortment of biomacromolecules;

- how these biomacromolecules in turn could arrange themselves into a hierarchy of functionally integrated systems; and

- how this spiraling of molecular complexity could ultimately produce the DNA-RNA-protein machinery required to perform the specific functions that we associate with actual living cells.

Of course, one can imagine a simple molecular replicator, by natural selection acting over many generations, gradually producing the functionally integrated multimolecular systems that we now associate with cellular life. But imagination is cheap. Where is the evidence to support such a scenario? All experiments that attempt to simulate the production of biologically-relevant building blocks under realistic prebiotic conditions are subject to interfering cross-reactions between desired chemical products and undesired chemical byproducts (see chapter 9). These reactions produce biologically irrelevant tars and melanoids that do not evolve in biologically productive directions. Accordingly, what nature selects on its own, in plausible prebiotic chemical environments (and thus without extensive investigator interference), tends to be chemically inert, biologically irrelevant, and therefore a dead-end to further chemical evolution.

To be sure, origin-of-life researchers are able to circumvent the problem of interfering cross-reactions by carefully designing experiments and extensively constraining or manipulating their outcomes. For instance, researchers can simply remove undesirable chemical reaction products or use purified starting materials. The problem of interfering cross-reactions is therefore averted, but at the cost of a new problem: such actions involve input of information by design and correspond to no known material process.

The work of Julius Rebek is emblematic of the problems facing molecular Darwinism. Rebek, while still at MIT before joining Scripps, synthesized amino adenosine triacid ester (AATE), which consists of two components, pentafluorophenyl ester and amino adenosine.[3] Moreover, he placed these AATE molecules in an organic solvent that preserves rather than degrades them. Both his choice of solvent and molecules represent an artificial constraint on the environment that has no natural analogue.

Even granting this artificial set-up, the evolution of Rebek's molecules hardly describes a plausible route for life's origin. To be sure, Rebek's molecules make copies of themselves and therefore replicate. But, as Gerald Joyce pointed out, they reproduce too accurately.[4] Without a sufficient amount of random variation, the Darwinian mechanism has nothing to select. In the extreme case, where things just keep producing identical copies of themselves, no evolution at all is possible. Still more problematic for Rebek's model is this: not only do Rebek's molecules carry far less information than is required for life, but they suggest no plausible pathway to the vast increase in information required for life.[5] Leslie Orgel sums up the significance of Rebek's work for the origin of life as follows: "What Rebek has done is very clever, but I don't see its relevance to the origin of life."[6]

For molecular Darwinism to succeed as an explanation of life's origin, it must explain why evolution proceeds in a *com-*

plexity-increasing direction. Starting with a simple replicator that could have emerged from a self-organizing world, molecular Darwinism needs to account for its evolution toward ever increasing complexity. Yet ironically, molecular Darwinism suggests that evolution should proceed in a *complexity-decreasing* direction. Consider, for instance, Sol Spiegelman's work on the evolution of polynucleotides in a replicase environment.[7] Leaving aside that the replicase protein is supplied by the investigator (from a viral genome), as are the activated mononucleotides needed to feed polynucleotide synthesis, the problem here, and in experiments like it, is the steady diminution of biologically relevant information over the course of the experiment. As Brian Goodwin notes:

> In a classic experiment, Spiegelman in 1967 showed what happens to a molecular replicating system in a test tube, without any cellular organization around it. The replicating molecules (the nucleic acid templates) require an energy source, building blocks (i.e., nucleotide bases), and an enzyme to help the polymerization process that is involved in self-copying of the templates. Then away it goes, making more copies of the specific nucleotide sequences that define the initial templates. But the interesting result was that these initial templates did not stay the same; they were not accurately copied. They got shorter and shorter until they reached the minimal size compatible with the sequence retaining self-copying properties. And as they got shorter, the copying process went faster. So what happened with natural selection in a test tube: the shorter templates that copied themselves faster became more numerous, while the larger ones were gradually eliminated. This looks like Darwinian evolution in a test tube. But the interesting result was that this evolution went one way: toward greater simplicity. Actual evolution tends to go toward greater complexity, species becoming more elaborate in their structure and behavior, though the process can also go in reverse, toward simplicity. But DNA on its own can go nowhere but toward greater simplicity. In order for the evolution of complexity to occur, DNA has to be within a cellular context; the whole system evolves as a reproducing unit.[8]

There are costs to becoming complex. For something to replicate, it has to reproduce itself in all essential respects. The bigger it is and the more it does, the greater the burden of replication. Simpler replicators therefore have an advantage over more complicated ones—they have less to keep track of in replication. Moreover, as in the Spiegelman case, natural selection is able to exploit this advantage. Darwinism places no inherent premium on complexity. In fact, other things being equal, the Darwinian mechanism prefers simplicity over complexity.[9] To be sure, other things are not always equal. In the history of life, where cells already possess vast amounts of complexity and vast behavioral repertoires, complexity has clearly increased. But there is no evidence that the Darwinian mechanism, when applied to individual molecules or simple molecular systems, accounts for evolution in a complexity-increasing direction. Molecular Darwinism therefore fails to resolve the problem of life's origin.

Theodosius Dobzhansky, a key architect of the neo-Darwinian synthesis, remarked that "prebiological natural selection is a contradiction in terms."[10] Dobzhansky may have overstated things, but not by much. It is possible to apply the Darwinian mechanism of random variation and natural selection at the molecular level to simple self-replicating systems. These systems will evolve provided they don't replicate too exactly and therefore allow some variability of offspring. Yet, in every instance, these simple molecular replicators have proven highly contrived and vastly simpler than actual biological organisms (cellular life). Moreover, they show no sign of evolving into anything that remotely resembles actual biological organisms—not in complexity and not in variety of functions. The quasi-mystical view that the Darwinian mechanism is a ratchet that can take just about any replicator and transform it into something vastly more splendid and complex has neither evidence nor theoretical foundation.

When All Else Fails—Panspermia

According to some origin-of-life researchers, the conditions on the early Earth were so inhospitable that life did not originate here. Rather, it originated elsewhere in the universe and then was seeded to Earth from outer space. Theories in which life comes to Earth by being seeded from outer space are known as *panspermia theories,* and they come in two forms.

In one form of the theory, spores that are able to withstand harsh conditions travel through space on dust or asteroids, land on Earth, and thereby seed it with the first life.[1] Once life is here, the Darwinian mechanism is supposed to kick in and life is supposed to evolve. There are serious difficulties with this proposal. Out of all the diversity of life forms, few, if any, could withstand the radiation or extremes of heat and cold found in space for periods like those necessary for transport between solar systems. Moreover, the distances between stars are immense. It seems unlikely that spores released near one star would be intercepted by a planet orbiting another.

To circumvent these difficulties, Francis Crick proposed a modification of the panspermia idea known as *directed panspermia:* intelligent aliens who travel in spaceships come to Earth and seed it intentionally with life (in an interstellar version of "Johnny Appleseed").[2] The spaceships protect the life forms and thus circumvent the difficulties associated with the "undirected" panspermia theories.

There are two serious difficulties with such theories. One is evidential: there simply is no evidence that life on Earth was seeded from outer space. Viable spores hitching a ride on asteroids and the like have not been found. Moreover, there is not a shred of evidence that intelligent aliens exist, much less that they came to Earth to seed it with life. The other serious difficulty is conceptual: both undirected and directed panspermia theories explain only the appearance of life on Earth, but not how life originated in the first place. In other words, they pass the buck.

The Medium and the Message

Origin-of-life researchers readily admit that they don't know how life actually got started. At the same time, many are quite confident that they know the broad contours for how life *could have* gotten started and that eventually they will figure out the details. In other words, they see themselves as reverse engineers. For reverse engineers, it's not crucial to follow the exact path by which an item was first produced. Rather, it is enough to find at least one plausible path by which it could have been produced. For origin-of-life researchers, it is therefore enough merely to establish a "proof of principle" or "proof of concept."[1] How life actually originated, though no doubt interesting, is largely irrelevant.

To appreciate that this is how origin-of-life researchers actually do view the problem of life's origin, consider the following two remarks by two well known origin-of-life researchers. The first is by Stuart Kauffman, the second by Leslie Orgel:

> Anyone who tells you that he or she knows how life started on the earth some 3.45 billion years ago is a fool or a knave. Nobody knows. Indeed, we may never recover the actual historical sequence of molecular events that led to the first self-reproducing, evolving molecular systems to flower forth more than 3 million millennia ago. But if the historical pathway should forever remain hidden, we can still develop bodies of theory and experiment to show how life might realistically have crystallized, rooted, then covered the globe. Yet the caveat: nobody knows.[2]

"Anybody who thinks they know the solution to this problem [of the origin of life][3] is deluded," says Orgel. "But," he adds, "anybody who thinks this is an insoluble problem is also deluded." One possible approach to the problem of life's origins is to ask the question scientifically rather than historically— how *can* life emerge rather than how *did* life emerge. In order to address this, scientists try to determine experimentally what is chemically feasible and what could have occurred on the prebiotic earth.[4]

These quotations oversell the prospects for resolving the problem of life's origin in purely materialist terms. Given the vast array of obstacles facing a purely material origin of life— obstacles outlined in detail in the previous chapters—why should anyone think that the problem of life's origin is soluble in terms of "what is chemically feasible and what could have occurred on the prebiotic earth"? Is there any evidence for this view? Or has it merely become an article of faith among origin-of-life researchers? In fact, the obstacles cited (e.g., racemic mixtures, interfering cross-reactions, the synthesis of polymers) suggest anything but a purely material origin of life. There is no evidence that material processes are able to overcome these obstacles. Moreover, to say that they must have this ability—because, after all, here we are—simply begs the question. The belief that material processes can overcome these obstacles is therefore without basis and indeed an article of faith.

To ascribe a purely material origin to life is to affirm that life is, without remainder, chemistry. Harvard chemist George Whitesides spoke on this very point to and for the 160,000-member American Chemical Society when in 2007 he received the society's highest honor, the annually awarded Priestley Medal. As part of this honor, Whitesides delivered an award address in which he discussed the origin of life: "This problem [of life's origin] is one of the big ones in science. It begins to place life, and us, in the universe. Most chemists believe, as do I, that life emerged spontaneously from mixtures of molecules in the prebiotic Earth. How? I have no idea."[5] Even so, Whitesides added that if anyone is going to resolve this problem, it will be the chemists: "I *believe* that understanding the cell is ultimately a question of chemistry and that chemists are, in principle, best qualified to solve it. The cell is a bag—a bag containing smaller bags and helpfully organizing spaghetti—filled with a Jell-O of reacting chemicals and somehow able to replicate itself."[6] In this way Whitesides reduced the problem of life's origin to a problem in chemistry.

Such a reduction of life, however, is fundamentally misconceived. Chemistry provides the medium for life, but the information that this medium carries, if life is to exist at all, is a message that cannot be reduced to it. "Obviously, life is a chemical phenomenon," writes Paul Davies, "but its distinctiveness lies not in the chemistry as such. The secret of life comes instead from its informational properties; a living organism is a complex information-processing system."[7] The medium of chemistry carries the message of life, but the medium cannot generate the message—the information. To suggest otherwise is like saying that ink and paper have the power to organize themselves into pages of meaningful text.[8] They have no such power.

Meaningful texts owe their meaning not to the physics and chemistry of ink on paper but to the infusion of semantic information. Likewise, life owes its origin not to the physics and chemistry of life's basic building blocks but to the infusion of biologically significant functional information. Whitesides is an outstanding chemist. He goes too far, however, in seeing everything as a problem of chemistry. He therefore misses that the origin of information is not a problem of chemistry. Chemistry can be a carrier of information, but it cannot be its source. Whitesides's failure to see this is perhaps not surprising. Chemists typically do not concern themselves with the problem of the origin of information because their work presupposes a smart chemist ready to provide it! But smart chemists infusing information into their chemical experiments are intelligent designers. Such experiments, if they support the origin of life at all, support not a materialist but an intelligent origin of life.

Biologists now recognize the crucial importance of information to understanding life and, especially, its origin. Caltech president and Nobel Prize-winning biologist David Baltimore, in describing the significance of the Human Genome Project, stated, "Modern biology is a science of information."[9] Origin-of-life researcher and fellow Nobel Prize winner Manfred Eigen has identified the problem of life's origin with the task of uncovering "the origin of information."[10] Biologists John Maynard Smith and Eörs Szathmáry have explicitly placed information at the center of developmental and evolutionary biology: "A central idea in contemporary biology is that of information. Developmental biology can be seen as the study of how information in the genome is translated into adult structure, and evolutionary biology of how the information came to be there in the first place."[11]

Given the importance of information to biology, the obvious question is, How does biological information arise? What is its source? Whitesides sees the ultimate source of biological information as residing in chemistry. Nobel laureate and origin-of-life researcher Christian de Duve develops this point more fully. In *Vital Dust,* he lays out seven "ages" in the history of life. Only the first four concern us in this chapter. Note the order: The Age of Chemistry, The Age of Information, The Age of the Protocell, and The Age of the Single Cell.[12] Here is how de Duve describes the transition from the first to the second age:

> History is a continuous process that we divide, in retrospect, into ages—the Stone Age, the Bronze Age, the Iron Age— each characterized by a major innovation added to previous accomplishments. This is true also of the history of life. . . . First, there is the Age of Chemistry. It covers the formation of a number of major constituents of life, up to the first nucleic acids, and is ruled entirely by the universal principles that govern the behavior of atoms and molecules. Then comes the Age of Information, thanks to the development of special information-bearing molecules that inaugurated the new processes of Darwinian evolution and natural selection particular to the living world.[13]

De Duve is here claiming that the ordering of nucleotides marks the advent of The Age of Information and that universal chemical principles govern everything in The Age of Chemistry leading up to it. But how does he know that such chemical principles provide a suitable and sufficient backdrop for the origin of biological information? Chemical principles describe both local law-like interactions among particles and quantum mechanical transitions among particle states. Nothing about such principles, however, guarantees the formation of nucleotide sequences or any other information-bearing molecules. Indeed, the principles of chemistry, though obviously compatible with the existence of biological information, offer no positive theoretical grounds for its origin.

Empirical evidence likewise fails to confirm de Duve's "chemistry first" view of life's origin. In particular, it fails to confirm that chemical processes operating under realistic prebiotic conditions can bring about nucleotide sequences or other information-bearing molecules. De Duve's "thioester world" is, in this respect, as ill-supported as its competitors (see chapter 16). For it to be well supported, Duve would need to provide a reasonably probable, fully articulated chemical pathway from his thioester world to the world of nucleotide sequences. He has accomplished nothing of the sort. If he had, prebiotic chemistry and chemical evolution would be going concerns and this book would reflect the vitality of materialist origin-of-life research.

To dispassionate eyes, materialist origin-of-life research is moribund. It seems unable to recapture the initial excitement of the Miller-Urey experiment over half a century ago. Those experiments were supposed to lay the groundwork for life's origin by supplying its basic building blocks. Yet those building blocks now seem incapable of spontaneously arranging themselves into information-rich biological structures. In short, the medium of chemistry gives no evidence of being able to generate the message of life.

The God of the Gaps

If chemistry alone is unable to account for the transition from The Age of Chemistry to The Age of Information, what, then, could account for it? With no empirical evidence or theoretical basis for the ability of chemistry by itself to generate biological information, one might think that the origin-of-life research community should be open to a design hypothesis that posits an information source capable of accounting for information-rich biological structures. But one would be mistaken. In fact, origin-of-life researchers, and biologists generally, tend to reject this option, charging those who defend an intelligent origin of life with committing a god-of-the-gaps fallacy (i.e., the fallacy of introducing a nonmaterial cause where a material cause, though for the present unknown, will do).

For instance, in his book *The Language of God,* Francis Collins, head of the Human Genome Project, admits that "no serious scientist would currently claim that a naturalistic explanation for the origin of life is at hand." Yet he immediately adds, "that is true today, and it may not be true tomorrow." Collins worries that an information source not reducible to de Duve's universal chemical principles is simply a place-holder for ignorance (i.e., a god of the gaps). Such a source could be assumed to exist only so long as, in Collins's words, "scientific understanding is currently lacking." As further justification for his concerns, Collins cites the history of science: "From solar eclipses in olden times to the movement of the planets in the Middle Ages, to the origins of life today, this 'God of the gaps' approach . . . may be headed for crisis if advances in science subsequently fill those gaps."[1]

When Collins discusses the origin of life elsewhere, however, he seems less concerned about the god of the gaps and more open to a design hypothesis: "Four billion years ago, the conditions on this planet were completely inhospitable to life as we know it; 3.85 billion years ago, life was teeming. That is a very short period—150 million years—for the assembly of macromolecules into a self-replicating form. I think even the most bold and optimistic proposals for the origin of life fall well short of achieving any real probability for that kind of event having occurred." Collins then asks whether the origin of life might be where a nonmaterial intelligence entered natural history. In answer, he says, "I am happy to accept that model." This is a remarkable concession. To be sure, he immediately qualifies it by warning against the god of the gaps. But he then qualifies this qualification by further highlighting the sheer magnitude of the origin-of-life problem: "this particular area of evolution, the earliest step, is still very much in disarray."[2]

Origin-of-life researcher Robert Hazen likewise worries that to posit an information source not reducible to chemistry commits a god-of-the-gaps fallacy. According to him, proponents of intelligent design argue that "life is so incredibly complex and intricate that it must have been engineered by a higher being." This in turn means that "no random natural process could possibly lead from non-life to even the simplest cell, much less humans." The problem for Hazen with this argument is that hypothesizing an information source not reducible to chemistry "ignores the power of emergence to transform natural systems without conscious intervention." But appealing to emergence, as we saw in chapter 16, is itself a stop-gap for ignorance. Hazen tacitly admits as much when, right after invoking emergence as the general solution to life's origin, he concedes, "True, we don't yet know all the details of life's genesis story, but why resort to an unknowable alien intelligence when natural laws appear to be sufficient?"[3]

Hazen's question contains two false assumptions: the sufficiency of natural laws and the unknowability of the designer. The claim that natural laws are sufficient to account for the origin of life is far-fetched. Natural laws work *against* the origin of life. Natural laws describe material processes that *consume* the raw materials of life, turning them into tars, melanoids, and other nonbiological substances that thereafter are completely useless to life. For the raw materials of life to avoid being consumed in this way, material processes must, as this book has documented in painstaking detail, overcome a daunting set of obstacles (see chapters 7 through 13). Yet material processes give no evidence of being able to overcome these obstacles. Rather, they give evidence of being predisposed to crash into them and never to get past them. Life has clear needs, and natural laws, far from supporting or even being neutral about those needs, directly undercuts them.

Hazen is also mistaken that any intelligence able to account for the information in biological systems must be "unknowable." We know intelligences through what they do, that is, by attending to the objects and events they produce. Knowledge of intelligences gained by examining their products may be limited, but it can be accurate as far as it goes. Thus, if intelligence is responsible for the origin of life, the study of biological systems yields knowledge about it. At the very least, we can know that such an intelligence is highly skilled in nano-engineering. This is not to say that the intelligence behind biology is merely a nano-engineer. But it is not less than a nano-engineer.

The worry that some presently unknown materialist explanation will one day dawn on scientists and thereby invalidate a design explanation of life's origin seems therefore overblown. Science is a bold enterprise. It takes risks and can afford to take risks because it is always in contact with empirical evidence and therefore can correct itself in light of new facts. The design hypothesis for the origin of life may ultimately be shown wrong, but so what? All scientific hypotheses place themselves in empirical harm's way and may be shown wrong.

Thus, if a compelling materialist account for the origin of life is one day found, it will overthrow the design hypothesis by rendering it superfluous (note that this very possibility shows that intelligent design is testable). On the other hand, the design hypothesis may well be right. Yet if it is, the only way to determine its rightness is by admitting the hypothesis as a live scientific option and then subjecting it to the severe scrutiny characteristic of science. As things stand, by rejecting the design hypothesis out of hand, origin-of-life researchers artificially prop up materialist origin-of-life scenarios, withholding from them the scrutiny they deserve.

If appealing to unknown materialist explanations is not a good reason for rejecting design, then neither is Francis Collins's appeal to the history of science. Collins compares invoking design to explain life's origin with invoking design to explain eclipses and the motion of planets. But when in times past people invoked the action of an intelligence to explain eclipses or the motion of planets, it was in ignorance of the relevant astronomical facts underlying these phenomena. We find ourselves in a radically different situation with regard to life's origin: now that we know the relevant facts of biochemistry and molecular biology, we are in a position to assess how difficult it is for the chemical building blocks of life to arise and then arrange themselves into the information-rich structures required for cellular life. So long as design hypotheses are based on knowledge rather than ignorance, they are scientifically legitimate.

Ironically, the history of science suggests that design explanations of life's origin may be *less* problematic than materialist explanations. In Darwin's day, the cell was regarded as extremely simple—it was thought to be essentially a blob of jelly enclosed by a membrane. Given this apparent simplicity, the origin of cellular life did not seem to require the long gradual ascent up a hierarchy of complexity as with Oparin's hypothesis. Instead, cells were thought to form instantaneously from pre-organic matter (see the discussion of "spontaneous generation" in chapter 3). Instantaneous spontaneous generation strikes us now as ludicrous given what biochemistry and molecular biology reveal about the complexities of the cell. But Darwin and his nineteenth-century disciples, in ignorance of the cell's true complexities, mistook it for a simple structure and, as a consequence, attributed it to purely material factors, thereby eliminating design from the scientific discussion of life's origin. The history of science therefore reveals that uncritical acceptance of materialist explanations over design explanations can itself constitute an argument from ignorance. In fact, we might say that it commits an evolution-of-the-gaps fallacy![4]

Cellular Engineering

What, then, are the right reasons, if any, for accepting a design explanation of life's origin? Reasons for accepting a claim come in two forms, positive and negative, and it's always good to have both. For instance, in answering a multiple-choice test, it is best to have not just negative reasons for rejecting the incorrect answers but also positive reasons for accepting the correct answers. The same holds in science. With the origin of life, we have compelling negative reasons for rejecting a materialist explanation. These include

- the complete absence of plausible chemical pathways from realistic prebiotic scenarios to the information-rich structures found in cells;

- a catalogue of obstacles that any such pathways must overcome, that they display no ability to overcome, and that derail prebiotic chemistry before it can even get close to a materialist origin of life (see chapters 7 through 13); and

- the failure of self-organizational, Darwinian, and other materialist principles to provide a coherent theoretical framework for the origin of life.

In short, neither evidence nor theory supports a materialist origin of life.

Besides negative reasons for rejecting a materialist origin of life, there are also positive reasons for accepting an intelligent origin of life. These include

- the engineering features of cellular systems;

- the irreducibility of cellular systems; and

- a conservation principle from the mathematical theory of information showing that evolutionary processes must always take in at least as much information as they give out.

We examine these positive reasons in this and the next two chapters. We begin with cellular engineering.

Many of the systems inside the cell represent nanotechnology at a scale and sophistication that dwarfs human engineering. Moreover, our ability to understand the structure and function of these systems depends directly on our facility with engineering principles (both in developing the instrumentation to study these systems and in analyzing what they do). Engineers have developed these principles by designing systems of their own, albeit much cruder than what we find inside the cell. Many of these cellular systems are literally machines: electro-mechanical machines, information-processing machines, signal-transduction machines, communication and transportation machines, etc. They are not just analogous to humanly built machines but, as mathematicians would say, *isomorphic* to them, that is, they capture all the essential features of machines.[1]

The genetic-coding and protein-synthesizing machinery in cells is of particular interest for the origin of life. The chemical properties of amino acids, nucleotide bases, sugars, phosphates, etc. that make up this machinery are, in themselves, not adequate to build this machinery. In fact, the natural chemical tendencies of these compounds under realistic prebiotic conditions would have inhibited the formation of coded information, which is at the heart of this machinery. For instance, amino acids react with sugars, preventing the formation of DNA and RNA. To be sure, prebiotic chemistry is compatible with certain forms of order (repeating patterns, symmetry, etc.). Yet nowhere in nonbiological nature do we find coded information.

The coded information inside cells is mathematically identical (isomorphic) with the coded information in written language.[2] Since both written language and DNA have that telltale property of encapsulating information in specific sequences of characters, and since intelligence alone is known to produce written language, and since there is no known materialist route to the coded information inside cells, it is an entirely reasonable inference to identify the cause of the coded information in DNA with an intelligent information source. More generally, we know only one cause capable of producing such high-tech machinery as exists in all cells, namely, intelligence. The engineering features of cells therefore count as a positive reason to accept intelligent design.

This is not to say, however, that cells are nothing but fancy engineering. In emphasizing that cells employ molecular machines and that engineering principles are indispensable to understanding cells, design theorists are often accused of reducing life to mechanism, conceiving of living forms as machines rather than as organisms.[3] This charge commits a *fallacy of composition,* arguing incorrectly that what is true of the parts must be true of the whole.[4] Just because cells have machine-like aspects does not mean that they are machines. Indeed, the intelligent design community regards living forms as much more than machines.[5] But because life's machine-like aspects clearly signal the action of intelligence, they receive special attention in the theory of life's intelligent design.

Conversely, in attempting, under realistic prebiotic conditions, to reproduce various components of cells (e.g., RNA sequences, as in the RNA world, or metabolic cycles, as in metabolism-first worlds), origin-of-life researchers may be tempted to, and sometimes in fact do, commit a fallacy of division, arguing incorrectly that what is true of the whole must be true of the part.[6] Thus, they ascribe life to these components when in fact this ascription is undeserved and properly belongs to life taken as a whole. For instance, complexity theorist

Norman Packard founded a start-up company called ProtoLife with the goal of synthesizing life from non-living materials.[7] The complex organic structures that this company actually endeavors to synthesize, however, fall far short of cells and do not deserve to be called "life" or, for that matter, the equally misleading "protolife."

Interestingly, ProtoLife describes its research as belonging to "the field of complex chemical system *design*."[8] Its actual research program is therefore more consistent with a design-theoretic as opposed to materialist approach to life's origin.

Irreducible Complexity

A system is said to be irreducibly complex if simplifying it by eliminating parts means that key functions of the system are irretrievably lost.[1] With irreducible complexity, simplification doesn't just make things operate worse—it breaks them so that they don't operate at all. Cells are chock-full of irreducible complexity.[2] For instance, the protein synthesizing machinery inside the cell is not just irreducibly complex but forms a nested hierarchy of irreducibly complex systems (i.e., it consists of irreducibly complex systems within irreducibly complex systems).

Take a key component of this system, the ribosome. Ribosomes, which stitch amino acids into proteins by reading information from an RNA template, are immensely complex, requiring at minimum fifty or so proteins and RNAs. The ribosome contains a set of core constituents that are irreducibly complex. Moreover, it is part of a larger system of enzymes and polynucleotides that read genetic information off a DNA template to produce proteins. This larger system, which includes the ribosome, the sixty enzymes discussed in chapter 12, and more—each indispensable to all known cellular life—is itself irreducibly complex. Indeed, the cell is filled with systems and subsystems that are indispensable to cellular life and that are, in their own right, each irreducibly complex.

Some scientists argue that irreducible complexity is not really necessary because an earlier, simpler "precursor" system could have existed first and then later evolved into the more complex system. But if the later system is the protein synthesizing machinery, what earlier system could have evolved into it? Indeed, where is the precursor to the cell's protein synthesizing machinery that could function on its own in the absence of that machinery? Some of the irreducibly complex systems inside cells are dispensable in the sense that cells can get by without them. But no instance of cellular life is known that can "get by" without the protein synthesizing machinery. Because these systems are indispensable to cellular life as such, facile appeals to supposed evolutionary precursors will not do.

The only life we know is protein-based and presupposes an elaborate nested hierarchy of irreducibly complex machinery that synthesizes proteins. If this machinery evolved from a precursor that did something else (i.e., that did something other than build proteins), then either something else inside cells was building the proteins or the precursor was not in fact a living cell. In either case, we haven't the slightest idea what such a precursor might have been. Irreducible complexity of systems that are indispensable for life is therefore irreducible complexity on steroids. It rules out all appeals to precursors based simply on general evolutionary principles.

Short of a concrete proposal for how these systems might have evolved from simpler precursors that do not merely exhibit some function or other but also maintain viability of the cell, evolutionary biology offers no clue, much less evidence, for their evolution. The fact is that no plausible chemical pathways to these systems from realistic prebiotic scenarios have been discovered—nothing even close. We therefore have no evidence that these systems are within the scope of materialist processes. On the other hand, we do know, as in the previous chapter on cellular engineering, that intelligence is able to produce irreducibly complex systems. Accordingly, we have here a good positive reason for thinking that these systems are actually designed.

23

Conservation of Information

Many searches in science are needle-in-the-haystack problems, which look for small targets in large spaces. In such cases, blind search stands no hope of success. Success, instead, requires an assisted search. This assistance takes the form of information: searches require information to succeed. Think of an Easter egg hunt in which saying "warmer" or "colder" indicates that one is getting respectively closer to or farther from the eggs. Clearly, this additional information assists those who are searching for the Easter eggs, especially when the eggs are well hidden and blind search would be unlikely to find them. But how does one secure the information required for a search to be successful? To pose the question this way suggests that successful searches do not magically materialize but need themselves to be discovered by a process of search.

The question then naturally arises whether such a higher-level "search for a search" is ever easier than the original search. Work in the field of *evolutionary informatics* indicates that this is not the case. Evolutionary informatics, a branch of information theory that studies the informational requirements of evolutionary processes, shows that the information required to make a search successful obeys a conservation principle known as the *conservation of information*.[1] This principle states that the information required to locate a successful search is never less than the information required for the original search to be successful. In consequence, the higher-level search for a search is never easier than the original lower-level search.

To see what is at stake with conservation of information, imagine that you are on an island with buried treasure. The island is so large that a random search is highly unlikely to succeed in finding the treasure. Fortunately, you have a treasure map that will guide you to the treasure. But where did you get the treasure map? Treasure maps reside in a library of possible treasure maps. The vast majority of these will not lead to the treasure. How, then, did you happen to find the right map among all these possible treasure maps? What special information did you have to find the right one? Conservation of information says that the information needed to pick out the right map is never less than any information that enables you to locate the treasure directly.

Conservation of information admits a precise mathematical formulation.[2] It implies that information, like money or energy, is a commodity that obeys strict accounting principles. Just as corporations require money to power their enterprises and machines require energy to power their motions, so searches require information to power their success. Moreover, just as corporations need to balance their books and machines cannot output more energy than they take in, so searches, in successfully locating a target, cannot give out more information than they take in. Conservation of information has far reaching implications for evolutionary theory (for both chemical and biological evolution), highlighting that the success of evolutionary processes in exploring biological configuration space always depends on preexisting information. In particular, evolutionary processes cannot create the information required for their success from scratch.[3]

To get around this conclusion, evolutionary theorists sometimes deny that biological evolution constitutes a targeted search. For instance, Oxford biologist Richard Dawkins illustrates biological evolution with a computer simulation that employs a targeted search (the simulation is explicitly programmed to search for the target phrase METHINKS IT IS LIKE A WEASEL). But immediately after giving this illustration, he adds: "Life isn't like that. Evolution has no long-term goal. There is no long-distant target, no final perfection to serve as a criterion for selection."[4] Dawkins here fails to distinguish two equally valid and relevant ways of understanding targets: (1) targets as humanly constructed patterns that we arbitrarily impose on things to suit our interests and (2) targets as patterns that exist independently of us and therefore regardless of our interests. In other words, targets can be extrinsic (i.e., imposed on things from outside) or intrinsic (i.e., inherent in things as such).

In the field of evolutionary computing (to which Dawkins's METHINKS IT IS LIKE A WEASEL example belongs), targets are given extrinsically by programmers who attempt to solve problems of their choice and preference. But in biology, not only has life come about without our choice or preference, but there are only so many ways that matter can be configured to be alive and, once alive, only so many ways it can be configured to serve different biological functions. Most of the ways open to evolution (chemical or biological) are dead ends. Evolution may therefore be characterized as the search for the alternative "live ends." In other words, demographics—whatever facilitates survival and reproduction—sets the targets. Evolution, despite Dawkins's denials, is therefore a targeted search after all.

Because conservation of information shows that evolutionary processes must always take in at least as much information as they give out, Dawkins's attempt to see evolution as fundamentally a matter of building up complexity from simplicity cannot stand. For Dawkins, proper scientific explanation is "hierarchically reductionistic," by which he means that "a complex entity at any particular level in the hierarchy of organization" must be explained "in terms of entities only one level down the hierarchy."[5] Thus, according to Dawkins, "the one thing that makes evolution such a neat theory is that it explains how organized complexity can arise out of primeval simplicity."[6] This is also why Dawkins regards intelligent design as unacceptable: "To explain the origin of the DNA/protein machine by invoking a supernatural [*sic*] Designer is to explain precisely nothing, for it leaves unexplained the origin of the Designer. You have to say something like 'God was always there,' and if you allow yourself that kind of lazy way out, you might as well just say 'DNA was always there,' or 'Life was always there,' and be done with it."[7]

Conservation of information shows that Dawkins's primeval simplicity is not nearly as simple as he thinks. Indeed, what Dawkins regards as intelligent design's predicament of failing to explain complexity in terms of simplicity now confronts materialist theories of evolution as well. In *Climbing Mount Improbable,* Dawkins argues that biological structures that at first blush seem vastly improbable with respect to pure randomness (i.e., blind search) become quite probable once the appropriate evolutionary mechanism (i.e., assisted search) is factored in to revise the probabilities.[8] But this revision of probabilities just means that blind search has given way to an assisted search. And the information that enables this assisted search to be successful now needs itself to be explained. Moreover, by conservation of information, that information is no less than the information that the evolutionary mechanism adds in outperforming pure randomness.[9]

The conservation of information therefore points to an information source behind evolution that imparts at least as much information to the evolutionary process as this process in turn is capable of expressing. In consequence, such an information source

- cannot be reduced to materialist causes,

- suggests that we live in an informationally open universe, and

- may reasonably be regarded as intelligent.

The conservation of information therefore counts as a positive reason to accept intelligent design.

Thinking Outside the Box

Many origin-of-life researchers search exclusively for materialist solutions to the problem of life's origin. As a result, they artificially restrict their problem-solving abilities in the one field, if any, that requires maximal resources for solving problems. Precisely because the problem of life's origin is so difficult, the full range of theoretical options for resolving it needs to be on the table. As Paul Davies notes: "We are a very long way from comprehending the how [of life's origin]. This gulf in understanding is not merely ignorance about certain technical details, it is a major conceptual lacuna. . . . My personal belief, for what it is worth, is that a fully satisfactory theory of the origin of life demands some radically new ideas."[1]

Yet to entertain radically new ideas, one must think outside the box. And this, it seems, is where origin-of-life research leaves much to be desired. Wedded to an outdated materialist dogma that rejects design-based hypotheses out of hand, origin-of-life research has been spinning its wheels for the last fifty-plus years. Successful problem-solving requires two forms of ingenuity: (1) the ingenuity of selecting the appropriate reference frame within which to solve the problem and (2) the ingenuity of working adeptly within that frame to find an effective solution. It seems that where origin of life research has gone off course is in limiting itself to a materialist reference frame that consistently fails to provide fruitful insights into the problem of life's origin.

The two-fold ingenuity required for successful problem solving may be illustrated with a classic puzzle used to assess creativity and ingenuity. Consider nine dots arranged in the form of a square as follows:[2]

What is the minimum number of line segments needed to connect all nine dots if they are joined continuously?

Many people *assume* that the line segments joining the dots have to be confined to the square implicitly outlined by the dots. But this assumption is gratuitous—the statement of the problem says nothing about confining the line segments to this implicit square. Given this faulty assumption, one can connect the dots in no fewer than five continuous line segments. But once this assumption is abandoned, and the possibility of drawing line segments outside the implicit square is taken seriously, the solution becomes straightforward:

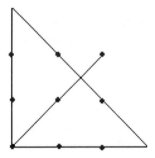

Thus we see that four continuously joined line segments are sufficient to connect the nine dots.

Expanding the reference frame for the problem of life's origin can do no harm and may actually do a lot of good in resolving this problem. In the nine-dots problem, moving to the unrestricted reference frame does not invalidate any of the candidate solutions proposed with respect to the more restrictive reference frame. It is simply that the restrictive frame offers fewer candidate solutions and, as it is, does not contain the actual solution. The lesson here for origin-of-life research is clear: expanding the range of permissible hypotheses to include design-based explanations of life's origin is not to rule out materialist approaches but rather to take away their monopoly so that other solutions, which might be better, have a fair chance to succeed.[3]

A Reasonable Hypothesis

Is the intelligent origin of life a reasonable hypothesis? This book has laid out strong negative reasons for rejecting a materialist origin of life as well as strong positive reasons for accepting an intelligent origin of life. Nonetheless, for the design hypothesis to seem reasonable to scientists, it needs a tie-in to the way scientists do their work. Charles Sanders Peirce, one of the founders of American pragmatist philosophy, held that for a difference to be a difference it has to make a difference.[1] Scientists reading this book might therefore wonder what practical difference accepting a design hypothesis might make to their understanding and investigation of life's origin. To be sure, chemistry by itself seems completely hopeless for generating information-bearing molecules, much less their hierarchical arrangement inside a cell. But what are we to make of an intelligence that is the source of biological information? If this information source is a supernatural agent that miraculously intervenes to produce biological information, then what hope is there of studying the origin of life scientifically?

To answer this question, note first that an intelligence that brought life into existence need not be supernatural—it could, as the ancient Stoics thought, be a natural feature of the natural world, thus making intelligence part of nature rather than beyond nature. Next, note that however an intelligence may have acted to originate life (whether by channeling ordinary material processes or by overriding them), design-theoretic research into life's origin can simply investigate the transformations of matter by which life *could* originate. Such an investigation is entirely parallel to what origin-of-life research already does, namely, attempt to show how life could originate and not necessarily how it did (recall chapter 19). The only difference is that whereas materialist research into life's origin is always supposed to restrict itself to "realistic prebiotic conditions" (and thus to chemistry undirected by intelligence), design-theoretic research into life's origin lifts that restriction.

Truth be told, origin-of-life research, as currently practiced, already lifts that restriction. Consider how philosopher of biology Michael Ruse summarizes research into the RNA world: "At this point, no one has actually confirmed that RNA can start self-replicating, nor has such an event been reproduced experimentally. . . . At the moment, the hand of *human design and intention* hangs heavily over everything, but work is going forward rapidly to create conditions in which molecules can make the right and needed steps without constant outside help."[2] If the hand of human design and intention indeed hangs heavily over *everything* in origin of life research, how can the work to show that "constant outside help" is unnecessary be "going forward rapidly"? The need for outside help in origin-of-life research suggests not a materialist but a designer-assisted origin of life. So long as origin-of-life research is constantly requiring outside help of a sort that would not have been available under "realistic prebiotic conditions," its progress in showing how life might have originated without outside help is nil.

Origin-of-life research promises with one side of the mouth to stay true to materialist principles but with the other asks indulgence for constantly needing to resort to the hand of human design and intention. Such equivocations are rampant throughout the field. Michael Ruse's comments about the RNA world are a case in point. Here is another: One of us (Wm. A. D) debated physicist James Trefil at Boston University in 2005 on the topic of intelligent design.[3] Trefil, who is associated with Harold Morowitz's origin-of-life research group at George Mason University (recall from chapter 16 that Morowitz works on metabolism-first worlds), claimed at the end of his presentation that he and his colleagues were but a few years from creating a life form in the lab.

To be sure, Trefil immediately qualified this claim by saying that such a life form would be quite different from life as we know it (i.e., it would not be a full-fledged cell). Yet the thing to note is that regardless of what Trefil and his associates create in the lab, it is *they* who will be doing the creating. That is, scientists will be in the lab using their intelligence to create something that they regard as alive. Whether what they come up with deserves to be called alive is here beside the point. The point is that by using their intelligence to do things that matter—left to its own devices—would never do, they are in fact engaged in design-theoretic research. For this reason, design theorist Paul Nelson recommends that origin-of-life researchers line their labs with mirrors to remind themselves of their role as intelligent designers in feeding their experiments with information and guiding them along paths that nature, operating under "realistic prebiotic conditions," would never have taken.

From a materialist perspective, the origin of life is a difficult problem because it is difficult to find the actual, necessary steps by which a blind materialist process could have created life. Find the right "chemical minestrone," as Stuart Kauffman says, and the origin-of-life problem goes away.[4] Materialist origin-of-life researchers therefore expect that by trying out enough recipes for such a chemical minestrone, they will eventually hit on the right one and unlock the mystery of life's origin. From a design-theoretic perspective, by contrast, the origin of life is a difficult problem because biological designs vastly exceed human designs in technological sophistication, and we are only beginning to grasp the technology. Design theorists therefore expect that by improving their understanding of technology, especially the nano-technology inside the cell, they will gain increasing insight into life's intelligent origin. At the same time, design theorists admit that certain features of the cell may so outstrip human understanding that a full grasp of life's intelligent origin may forever elude humanity. Regardless of who is right, given the evidence described in this book, the burden of proof has now shifted to the materialist origin-of-life researcher, who can no longer merely assert that cellular life arose by purely materialist processes but must now demonstrate in detail how it could have happened.

Will the origin-of-life community assume this burden of proof? Not until it is ready to question a prior commitment to materialism. Richard Dawkins is as ardent a materialist as one will find.[5] Yet he writes, "The illusion of purpose is so powerful that biologists themselves use the assumption of good design as a working tool."[6] Scientists are pragmatists and therefore loathe to give up effective tools that help them in their research. Design, ironically, is one such tool. Yet, conditioned by a materialist outlook that denies design, many scientists find it difficult to acknowledge design even when they are using it, imagining that it is merely an illusion.

We would therefore turn Dawkins's statement around: *Biologists use the assumption of design with such success as a working tool precisely because design in biology is not an illusion but real.*

Epilogue: Atheism as a Speculative Faith

To defeat an enemy is to defeat its strongest champion. Thus, when Israel defeated the Philistines, it was enough for David to defeat Goliath. After that, the Philistines fled. Atheism's enemy is theism, and the Goliath it faces is the problem of life's origin. An intellectually fulfilled atheism looks to science to defeat theism. But atheism's "David" is nowhere in sight; science is far from rescuing atheism.

Science poses many obstacles to atheism, and the greatest of these is the origin of life. Darwin sidestepped the problem, as we noted earlier. For Richard Dawkins therefore to invoke Darwin as underwriting an intellectually fulfilled atheism is absurd. Darwin's theory, even if overwhelmingly confirmed (which it is not), does nothing to settle the problem of life's origin.

Until science can show that physical processes operating under realistic prebiotic conditions can bring about full-fledged cells from nonliving materials, intellectual fulfillment remains an atheistic pipedream. As this book has documented in painstaking detail, physical processes give every evidence of actively working *against* the formation of the molecular structures and complexes needed for life. If intellectual fulfillment depends on scientific validation, atheism—despite its long and dreary history[1]—has now become a speculative faith.

Atheism is a belief with scientific pretensions but no scientific backing.[2] It promises freedom from superstition but is itself the slave to superstition. It is an ideology even more intolerant and demeaning than anything Dawkins attacks in *The God Delusion.*

Notes

Introduction: Atheism's Quest for Intellectual Fulfillment

1. Richard Dawkins, *The Blind Watchmaker: Why the Evidence of Evolution Reveals a Universe Without Design* (New York: Norton, 1987), 6.

2. William B. Provine, from a February 12, 1998 talk titled "Evolution: Free Will and Punishment and Meaning in Life," abstract available at http://eeb.bio.utk.edu/darwin/Archives/1998ProvineAbstract. htm (last accessed April 7, 2008). This was Provine's keynote address for the 1998 Darwin Day celebration at the University of Tennessee, Knoxville. The actual quote is from one of his slides.

3. Charles Darwin, Letter to Joseph Hooker (1871), in Francis Darwin, ed., *The Life and Letters of Charles Darwin*, in 3 volumes (London: John Murray, 1887), III:18.

4. Quoted from David L. Hull, *Darwin and His Critics* (Chicago: University of Chicago Press, 1973), 123–124.

5. Francisco J. Ayala, "Darwin's Revolution," in *Creative Evolution?!*, eds. J. H. Campbell and J. W. Schopf (Boston: Jones and Bartlett, 1994), 4. The subsection from which this quote is taken is titled "Darwin's Discovery: Design without Designer."

Chapter 1: The Problem of Life's Origin

1. George Polya, *How to Solve It*, 1st ed. (Princeton: Princeton University Press, 1945). This quote does not occur in the second edition, which was published in 1957. For this quote and others from the out-of-print 1945 edition, see http://www-gap.dcs.st-and.ac.uk/~history/ Quotations/Polya.html (last accessed November 30, 2006).

2. The following paragraphs, up to the bullet points, are adapted from Michael Denton, *Evolution: A Theory in Crisis* (Bethesda, MD: Adler & Adler, 1985), 328–329.

3. Harvard biologist Richard Lewontin in *The New York Review of Books* writes, "We take the side of science *in spite of* the patent absurdity of some of its constructs, *in spite of* its failure to fulfill many of its extravagant promises of health and life, *in spite of* the tolerance of the scientific community for unsubstantiated just-so stories, because we have a prior commitment, a commitment to materialism. It is not that the methods and institutions of science somehow compel us to accept a material explanation of the phenomenal world, but, on the contrary, that we are forced by our *a priori* adherence to material causes to create an apparatus of investigation and a set of concepts that produce material explanations, no matter how counterintuitive, no matter how mystifying to the uninitiated." Quoted in Richard Lewontin, "Billions and Billions of Demons" (review of Carl Sagan's *The Demon-Haunted World: Science as a Candle in the Dark*), *The New York Review of Books* (9 January 1997): 31. Lewontin is saying more here than that some scientific ideas are counterintuitive. Rather, he is saying that no evidence whatsoever can overturn science's commitment to materialism. But if that is the case, then the commitment to materialism in science is itself not scientific because scientific claims are always in contact with evidence and capable of being overturned in light of new evidence.

4. Origin of life researchers do not attempt to find out exactly what happened. That may be unknowable. What they seek is a plausible pathway. It is somewhat like Thor Heyerdahl crossing the ocean in a small craft, to demonstrate that early peoples could do so. That doesn't prove that they did what he did; it only proves that they could have done so. This point is developed in more detail in chapter 19.

5. Obligate parasites, such as viruses and bacterial endosymbionts, which cannot live independently of host organisms, do not pose a counterexample to this claim that the simplest cell requires hundreds of genes to handle its basic tasks of living. That's because these parasites require for their very existence cells that *can* live independently of host organisms, and such non-parasitic cells *do* require a full complement of genes and capabilities, which puts them well above the minimum complexity threshold described here. If you will, obligate parasites are just like other cells in requiring hundreds if not thousands of genes, only they don't keep them on hand but let other organisms maintain those genes for them. For the simplest bacterial endosymbiont at the time of this writing (i.e., *Carsonella ruddii* with a 160-kilobase genome), see the work of Nancy Moran at the Univer-

sity of Arizona, http://eebweb.arizona.edu/faculty/moran/research. htm (last accessed March 28, 2007).

6. Lynn Margulis has proposed that eukaryotic cells evolved by one prokaryotic cell swallowing another, with the swallowed cell then serving as the nucleus of the newly formed eukaryotic cell. This proposal remains speculative, however. It is not clear by what mechanisms a prokaryotic cell could engulf and assimilate another prokaryotic cell and thereafter work together symbiotically (phago-cytosis, by which eukaryotic cells ingest food particles, seems not to be the right model here). Moreover, there are many differences between eukaryotic and prokaryotic cells that are not explained by Margulis's model, including differences in cell division, cytoplasmic membranes, locomotor organelles, respiratory enzymes, and elec-tron transport chains. See Lynn Margulis and Dorion Sagan, *Acquiring Genomes: A Theory of the Origins of Species* (New York: Basic Books, 2002), ch. 3.

7. Karl Popper, "Scientific Reduction and the Essential Incomplete-ness of All Science," *Studies in the Philosophy of Biology* 259 (1974): 270. Emphasis in original.

Chapter 2: A Batch of Red Herrings

1. See http://astrobiology.arc.nasa.gov/about/index.cfm (last ac-cessed November 16, 2006).

2. The field of "artificial life," which had its heyday in the 1980s and 90s, and seems now to have run out of steam, emphasized evolv-ability of virtual organisms in virtual environments to the exclusion of the actual functional requirements of actual living systems. For artificial life's swan song, see Christoph Adami, *Introduction to Ar-tificial Life* (New York: Springer, 1999).

3. Robert M. Hazen, *Genesis: The Scientific Quest for Life's Origin* (Washington, DC: Joseph Henry Press, 2005), 189.

4. On a related note, life forms do not strictly need the ability to reproduce. The mule is recreated anew in each generation. However, the mule requires its parent species to continue to reproduce. Lack of procreative power means that the mule cannot have offspring, not that it cannot be alive.

5. Kalin Vetsigian, Carl Woese, and Nigel Goldenfeld, "Collec-tive Evolution and the Genetic Code," *Proceedings of the National Academy of Sciences* 103(28) (July 11, 2006): 10696–10701. W. Ford Doolittle and Eric Bapteste, "Pattern Pluralism and the Tree of

Life Hypothesis," *Proceedings of the National Academy of Sciences* 104(7) (February 13, 2007): 2043–2049.

Chapter 3: Spontaneous Generation

1. Thomas Kuhn, *The Structure of Scientific Revolutions*, 2nd ed. (Chicago: University of Chicago Press, 1970).

2. Besides Kuhn's *Structure of Scientific Revolutions* see also W. V. Quine and J. S. Ullian, *The Web of Belief*, 2nd ed. (New York: Random House, 1978).

3. Charles Darwin, Letter to Joseph Hooker (1871), in Francis Darwin, ed., *The Life and Letters of Charles Darwin*, in 3 volumes (London: John Murray, 1887), III:18.

4. Ernst Haeckel. *The Wonders of Life,* trans. J. McCabe (London: Watts, 1905), 111. Compare Thomas Henry Huxley, "On the Physical Basis of Life," *The Fortnightly Review* 5 (1869): 129–45.

5. Philip F. Rehbock, "Huxley, Haeckel, and the Oceanographers: The Case of Bathybius haeckelii," *Isis* 66(4) (1975): 504–533.

6. Christopher Wills and Jeffrey Bada, *The Spark of Life: Darwin and the Primeval Soup* (New York: Perseus, 2000), 24–25.

7. Richard Dawkins has remarked that "Darwin made it possible to be an intellectually fulfilled atheist." Yet without a materialist account of life's origin, one cannot be a completely intellectually fulfilled atheist. Richard Dawkins, *The Blind Watchmaker: Why the Evidence of Evolution Reveals a Universe without Design* (New York: Norton, 1987), 6.

Chapter 4: Oparin's Hypthesis

1. A. I. Oparin, *The Origin of Life on Earth*, 3rd ed (1924; reprinted, revised, and translated New York: Academic Press, 1957).

2. When Darwin wrote his *Origin of Species* in the mid 1800s, the cell was conceived as a blob of jelly enclosed by a membrane. Accordingly, it was thought to be so simple that it could have easily originated spontaneously and quickly. By the 1920s, when Oparin published his hypothesis suggesting that the first cell formed gradually, scientists understood the cell to be far more complex than they had previously imagined. They therefore recognized that without intelligent design the formation of something as complex as a cell must have required extended periods of time.

3. J. B. S Haldane, "The Origin of Life," *Rationalist Annual* 148 (1928): 3–10.

Chapter 6: Primitive Undersea Simulation Experiments

1. J. B. Corliss, J. A. Baross, and S. E. Hoffman, "An Hypothesis concerning the Relationship between Submarine Hot Springs and the Origin of Life on Earth," *Oceanologica Acta* 4(suppl.) (1981): 59–69.

2. C. Huber and G. Wächtershäuser, "Peptides by Activation of Amino Acids with CO on (Ni,Fe)S Surfaces: Implications for the Origin of Life," *Science* 281 (1998): 670–72.

3. Jay A. Brandes, Nabil Z. Boctor, George D. Cody, Benjamin A. Cooper, Robert M. Hazen, and Hatten S. Yoder Jr., "Abiotic Nitrogen Reduction on the Early Earth," *Nature* 395 (24 September 1998): 365.

4. Paul Davies makes this point as follows: "There would have been no lack of available energy sources on the early Earth to provide the work needed to forge [biologically significant chemical] bonds, but just throwing energy at the problem is no solution. The same energy sources that generate organic molecules also serve to destroy them. To work constructively, the energy has to be targeted at the specific reaction required. Uncontrolled energy input, such as simple heating, is far more likely to prove destructive than constructive." The same could be said for turbulence. Quoted from Paul Davies, *The Fifth Miracle: In Search for the Origin and Meaning of Life* (New York: Simon & Schuster, 1999), 89–90.

Chapter 7: Free Oxygen

1. J H. Carver, "Prebiotic Atmospheric Oxygen Levels," *Nature* 292 (1981): 136–38; James F. Kasting, "Earth's Early Atmosphere," *Science* 259 (1993): 920–26.

2. Philip H. Abelson, "Chemical Events on the Primitive Earth," *Proceedings of the National Academy of Sciences USA* 55 (1966): 1365–1372.

3. Marcel Florkin, "Ideas and Experiments in the Field of Prebiological Chemical Evolution," *Comprehensive Biochemistry* 29B (1975): 241–242.

4. Sidney W. Fox and Klaus Dose, *Molecular Evolution and the Origin of Life,* rev. ed. (New York: Marcel Dekker, 1977), 43, 74–76.

5. Jon Cohen, "Novel Center Seeks to Add Spark to Origins of Life," *Science* 270 (1995): 1925–1926.

6. Heinrich D. Holland, *The Chemical Evolution of the Atmosphere*

and Oceans.(Princeton: Princeton University Press, 1984), 99–100.

7. Gordon Schlesinger and Stanley L. Miller, "Prebiotic Synthesis in Atmospheres Containing CH4, CO, and CO2: I. Amino Acids," *Journal of Molecular Evolution* 19 (1983): 376.

8. John Horgan, "In the Beginning . . . ," *Scientific American* (February 1991): 116–126.

Chapter 9: Interfering Cross-Reactions

1. Alan Schwartz, editor of *Origins of Life and Evolution of Biospheres*, notes, "A problem which is familiar to organic chemists is the production of unwanted byproducts in synthetic reactions. For prebiotic chemistry, where the goal is often the simulation of conditions on the prebiotic Earth and the modeling of a spontaneous reaction, it is not surprising—but nevertheless frustrating—that the unwanted products may consume most of the starting material and lead to nothing more than an intractable mixture, or 'gunk.'" Quoted from Alan W. Schwartz, "Intractable Mixtures and the Origin of Life," *Chemistry & Biodiversity* 4(4) (2007): 656.

2. Charles Thaxton, Walter Bradley, and Roger Olsen, *The Mystery of Life's Origin: Reassessing Current Theories* (New York: Philosophical Library, 1984), 104–6.

3. Robert Shapiro, "A Simpler Origin of Life," *Scientific American* (February 12, 2007): available online at http://sciam.com/print_version.cfm?articleID=B7AABF35-E7F2-99DF-309B8CEF02B5C4D7 (last accessed April 13, 2007).

4. J. Brooks and G. Shaw, *Origin and Development of Living Systems* (New York: Academic Press, 1973), 359.

Chapter 10: Racemic Mixtures

1. Robert M. Hazen, Timothy R. Filley, and Glenn A. Goodfriend. "Selective Adsorption of L- and D-Amino Acids on Calcite: Implications for Biochemical Homochirality," *Proceedings of the National Academy of Sciences* 98 (8 May 2001): 5487–90.

2. Jessica Gorman, "Rocks May Have Given a Hand to Life," *Science News* 159(18) (5 May 2001), available online at http://www.sciencenews.org/20010505/fob1.asp (last accessed February 28, 2007).

Chapter 11: The Synthesis of Polymers

1. For simplicity, we limited ourselves in this analogy to linear arrangements of letters. Yet, because the bonding between amino-acid residues allows for branching, we could have extended the analogy

by not only rotating letters but also having them assume a branched pattern such as the following:

2. See William A. Dembski and Jonathan Wells, *The Design of Life: Discovering Signs of Intelligence in Biological Systems* (Dallas: Foundation for Thought and Ethics, 2008), see the sidebar "Monkeys Typing Shakespeare" in sec. 7.4.

3. J. Bowie and R. Sauer, "Identifying Determinants of Folding and Activity for a Protein of Unknown Sequences: Tolerance to Amino Acid Substitution," *Proceedings of the National Academy of Sciences* 86 (1989): 2152–56; J. Bowie, J. Reidhaar-Olson, W. Lim, and R. Sauer, "Deciphering the Message in Protein Sequences: Tolerance to Amino Acid Substitution," *Science* 247 (1990):1306–10. J. Reidhaar-Olson and R. Sauer, "Functionally Acceptable Solutions in Two Alpha-Helical Regions of Lambda Repressor," *Proteins, Structure, Function, and Genetics* 7 (1990): 306–10. See also Michael Behe, "Experimental Support for Regarding Functional Classes of Proteins to be Highly Isolated from Each Other," in *Darwinism: Science or Philosophy?*, eds. J. Buell, and G. Hearn (Dallas: Foundation for Thought and Ethics, 1994), 60–71; and Hubert Yockey, *Information Theory and Molecular Biology* (Cambridge: Cambridge University Press, 1992), 246–58.

4. Douglas D. Axe, "Extreme Functional Sensitivity to Conservative Amino Acid Changes on Enzyme Exteriors," *Journal of Molecular Biology* 301 (2000): 585–595.

5. The work of Sauer and Axe avoids the "retrospective fallacy" that design-critic Kenneth Miller attributes to intelligent design. When design theorists calculate the improbability of certain proteins, here's what they do according to Miller: "'It's what statisticians call a retrospective fallacy.' It is like equating the odds of drawing two pairs in poker with the odds of drawing a particular two-pair hand—say a pair of red queens, a pair of black 10s and the ace of clubs. 'By demanding a particular outcome, as opposed to a functional outcome, you

stack the odds,' Miller says. What these calculations fail to recognise is that many different protein sequences can be functional. It is not uncommon for proteins in different species to vary by 80 to 90 per cent, yet still perform the same function." (Bob Holmes and James Randerson, "A Skeptic's Guide to Intelligent Design," *New Scientist* 187 [July 9, 2005]: 11.) By contrast, Sauer and Axe have taken pains to avoid this fallacy, calculating not the improbability of a particular amino-acid sequence but the improbability of any sequence with the same fold and function. Note that the term "retrospective fallacy" is idiosyncratic: in fact, statisticians do not use this term to describe the mistake to which Miller is referring. What Miller is calling a fallacy is simply a failure to calculate the relevant probability.

6. "The Murchison meteorite contains a complex suite of amino acids. Whereas terrestrial samples are dominated by the 20 protein amino acids, over 70 amino acids have been positively identified in this meteorite, many of which appear to be uniquely extraterrestrial." S. Pizzarello and J. R. Cronin, "Alanine Enantiomers in the Murchison Meteorite," *Nature* 394(6690) (July 16, 1998): 236.

Chapter 13: Harnessing the Sun

1. Michael Denton, *Nature's Destiny* (New York: Free Press, 1998), 137.

2. "Competing hypotheses seek to explain the evolution of oxygenic and anoxygenic processes of photosynthesis. Since chlorophyll is less reduced and precedes bacteriochlorophyll on the modern biosynthetic pathway, it has been proposed that chlorophyll preceded bacteriochlorophyll in its evolution. However, recent analyses of nucleotide sequences that encode chlorophyll and bacteriochlorophyll biosynthetic enzymes appear to provide support for an alternative hypothesis. This is that the evolution of bacteriochlorophyll occurred earlier than the evolution of chlorophyll." Quoted from Peter J. Lockhart, Anthony W. D. Larkum, Michael A. Steel, Peter J. Waddell, and David Penny, "Evolution of Chlorophyll and Bacteriochlorophyll: The Problem of Invariant Sites in Sequence Analysis," *Proceedings of the National Academy of Sciences USA* 93(5) (1996): 1930–1934.

3. See, for instance, "Photosynthesis Analysis Shows Work of Ancient Genetic Engineering," *Science Daily* (November 22, 2002): available online at http://www.sciencedaily.com/releases/2002/11/021122074236.htm (last accessed April 16, 2007). The article summarizes an Arizona State University press release on

some work of Robert Blankenship. Merely by examining DNA sequences in extant bacteria, Blankenship and colleagues claimed to obtain "clear evidence that photosynthesis did not evolve through a linear path of steady change and growing complexity but through a merging of evolutionary lines that brought together independently evolving chemical systems—the swapping of blocks of genetic material among bacterial species known as horizontal gene transfer." Other than showing that *if* photosynthesis evolved, its evolutionary history must have been complicated, Blankenship's research does nothing to describe the actual path of evolution, much less to determine whether such an evolutionary path could have proceeded on purely materialist principles.

4. Compare the previous endnote with the following article, also involving work by Robert Blankenship, which indicates how much yet remains to be understood about the photosynthetic apparatus and its superb design capabilities: Gregory S. Engel, Tessa R. Calhoun, Elizabeth L. Read, Tae-Kyu Ahn, Tomãiš Man al, Yuan-Chung Cheng, Robert E. Blankenship, and Graham R. Fleming, "Evidence for Wavelike Energy Transfer through Quantum Coherence in Photosynthetic Systems," *Nature* 446 (2007): 782–786.

Chapter 14: The Proteinoid World

1. In saying that early-Earth history was more conducive toward the breakdown rather than the buildup of biological complexity, we are not backhandedly invoking the second law of thermodynamics. Instead, we are simply looking at the material processes that most likely were available on the early Earth and, by taking them on their own terms, inferring that they were unlikely to facilitate the buildup of biological complexity.

2. Thaxton *et al.*, *Mystery of Life's Origin*, 162.

3. *Ibid.*, 174–76.

Chapter 15: The RNA World

1. The authors of this book are in principle opposed to referring to DNA as "doing" anything at all. More accurate would be to say "the cell needs such-and-such in order to make copies of its DNA." Unfortunately, in the current state of biological discourse, it is almost impossible to say anything about these issues without begging many questions!

2. Gerald Joyce, "RNA Evolution and the Origins of Life," *Nature* 338 (1989): 217–24

3. Shapiro, "A Simpler Origin of Life." To illustrate the magnitude of the problem, Shapiro considers the analogy of a golfer "who having played a golf ball through an 18-hole course, then assumed that the ball could also play itself around the course in his absence. He had demonstrated the possibility of the event; it was only necessary to presume that some combination of natural forces (earthquakes, winds, tornadoes and floods, for example) could produce the same result, given enough time. No physical law need be broken for spontaneous RNA formation to happen, but the chances against it are so immense, that the suggestion implies that the non-living world had an innate desire to generate RNA. The majority of origin-of-life scientists who still support the RNA-first theory either accept this concept (implicitly, if not explicitly) or feel that the immensely unfavorable odds were simply overcome by good luck." Two comments are in order: (1) invoking "immense good luck" does not constitute a scientific theory and (2) the implicit small probability argument here suggests that such a formation of RNA would exhibit specified complexity and therefore trigger a design inference.

4. Joyce, "RNA Evolution and the Origins of Life."

5. Quoted in Robert Irion, "RNA Can't Take the Heat," *Science* 279 (1998): 1303.

6. See S. Klug and M. Famulok, "All You Wanted to Know about SELEX," *Molecular Biology Reports* 20 (1994): 97–107.

7. In "A Simpler Origin of Life," Shapiro considers the possibility of a "pre-RNA world" in which "the bases, the sugar or the entire backbone of RNA have been replaced by simpler substances, more accessible to prebiotic syntheses." Nevertheless, if origin-of-life researchers want such a nonnucleotide "soup" to produce a replicator, even a simple one, they will need to overcome some daunting probabilistic hurdles. According to Shapiro, "the spontaneous appearance of any such replicator without the assistance of a chemist faces implausibilities that dwarf those involved in the preparation of a mere nucleotide soup." Citing Gerald Joyce and Leslie Orgel for regarding it a "near miracle" that RNA chains on the lifeless Earth could form spontaneously, Shapiro concludes that it would similarly be a miracle for any of "the proposed RNA substitutes" to arrange themselves into a replicator.

Chapter 16: Self-Organizing Worlds

1. Ian Stewart, *Life's Other Secret: The New Mathematics of the Living World* (New York: John Wiley, 1998), 48.

2. More realistic assessments of the origin of life problem can be found in Thaxton *et al.*, *Mystery of Life's Origin* (1984); Robert Shapiro, *Origins, A Skeptics Guide to the Creation of Life on Earth* (New York: Summit Books, 1986); Paul Davies, *The Fifth Miracle* (1999); and Hubert Yockey, *Information Theory, Evolution, and the Origin of Life* (Cambridge: Cambridge University Press, 2005).

3. Harold J. Morowitz, *The Emergence of Everything: How the World Became Complex* (New York: Oxford University Press, 2002), 76. Emphasis added.

4. "Perhaps two-thirds of scientists publishing in the origin-of life field (as judged by a count of papers published in 2006 in the journal *Origins of Life and Evolution of the Biosphere*) still support the idea that life began with the spontaneous formation of RNA or a related self-copying molecule." Shapiro, "A Simpler Origin of Life." Throughout this article Shapiro refers to the "daunting probabilistic hurdles" that these genetics-first scenarios must overcome.

5. The sheer diversity of these explanations shows that origin-of-life theorizing is unconstrained by data, offers no insight into how life actually got started, and is limited only by our imagination. Now imagination may be a good thing, but it is too thin a soup on which to nourish science.

6. Stuart Kauffman, *At Home in the Universe: The Search for the Laws of Self-Organization and Complexity* (New York: Oxford University Press, 1995), 274.

7. Christian de Duve, *Singularities: Landmarks on the Pathways to Life* (Cambridge: Cambridge University Press, 2004).

8. Günter Wächtershäuser, "Evolution of the First Metabolic Cycles," *Proceedings of the National Academy of Sciences* 87 (1990): 200–204 and Günter Wächtershäuser, "Life as We Don't Know It," *Science* 289 (2000): 1307–1308.

9. Michael Russell, "Life from the Depths," *Science Spectra* 1 (1996): 26. See also William Martin and Michael Russell, (2002). "On the Origins of Cells: A Hypothesis for the Evolutionary Transitions from Abiotic Geochemistry to Chemoautotrophic Prokaryotes, and from Prokaryotes to Nucleated Cells," *Philosophical Transactions of the Royal Society: Biological Sciences* 358 (2002): 59–85.

10. David W. Deamer, "The First Living Systems: A Bioenergetic

Perspective," *Microbiology and Molecular Biology Reviews* 61 (1997): 239–261.

11. See Platts's website http://www.pahworld.com (last accessed April 20, 2007). For a popular exposition of Platts's views, see Robert M. Hazen, *Genesis: The Scientific Quest for Life's Origin* (Washington, D.C.: Joseph Henry Press, 2005), ch. 17.

12. Alexander G. Cairns-Smith, *Seven Clues to the Origin of Life* (Cambridge: Cambridge University Press, 1985); and Alexander G. Cairns-Smith and Hyman Hartman, eds., *Clay Minerals and the Origin of Life* (Cambridge: Cambridge University Press, 1986).

13. The situation here is like the police having seven possible suspects for a crime. They have one or at most two pieces of evidence that would implicate each one, but no pattern of evidence that would justify bringing charges against any of them. Stewart might say that things were going swimmingly because the police have "lots of suspects." By contrast, the police don't want "lots of suspects"; they want a good case against one suspect.

14. See Harold J. Morowitz, *Beginnings of Cellular Life: Metabolism Recapitulates Biogenesis* (New Haven: Yale University Press, 1992), chs. 10 and 12; Morowitz, *The Emergence of Everything*, 80.

15. Hazen, *Genesis*, 209–210.

16. *Ibid.*, 151. Emphasis added.

17. Wächtershäuser, like Morowitz, would like to reproduce such cycles under realistic prebiotic conditions. His iron-sulfur model, however, seems inadequate to the task. In "Self-Organizing Biochemical Cycles," *Proceedings of the National Academy of Sciences* 97(23) (2000): 12506, Leslie Orgel analyzes Wächtershäuser's model and argues that there is "no reason to expect that multistep cycles such as the reductive citric acid cycle will self-organize on the surface of FeS/FeS2 or some other mineral." Orgel immediately adds, "While it seems almost impossible that a cycle of reactions as complicated as the reductive citric acid cycle could self-organize on a mineral surface, Wächtershäuser's suggestion does raise an interesting and important question. How much self-organization is it reasonable to expect on a mineral surface in the absence of *evolved, informational catalysts*?" [Emphasis added.] Here is Orgel's answer: "It is not clear that any surface is likely to catalyze two or more unrelated chemical reactions." Nonspecialists may read this last statement as conceding that there is no solid empirical evidence that mineral surfaces can catalyze metabolic cycles—*in the absence of evolved, informational catalysts!*

18. Morowitz, *The Emergence of Everything*, 76.

19. "X emerges" is thus properly a shorthand for "X emerges from antecedent conditions Y via some method Z," where X, Y, and Z are precisely specified.

Chapter 17: Molecular Darwinism

1. We are indebted to Stephen Meyer for many of the insights in this chapter.

2. Throughout these discussions it is important to keep referring back to chapter 1, always bearing in mind that what needs to be explained is the cell as we know it and not the simplistic examples of life constantly put forward by origin-of-life researchers.

3. This work is summarized in Julius Rebek Jr., "Synthetic Self-Replicating Molecules," *Scientific American* 271(1) (1994): 48–55.

4. John Horgan, "In the Beginning," *Scientific American* 264(2) (Feb. 1991): 120.

5. See Lawrence Hurst and Richard Dawkins, "Life in a Test Tube," *Nature* 357 (21 May 1992):198–99.

6. Horgan, "In the Beginning," 120. In fact, Rebek's experiment makes a strong case for the tremendous power of a little design in overcoming random chemistry. In other words, it illustrates the power of intelligent design!

7. Sol Spiegelman, "An *In Vitro* Analysis of a Replicating Molecule," *American Scientist* 55 (1967): 221. See also Norman R. Pace and Sol Spiegelman, "*In Vitro* Synthesis of an Infectious Mutant RNA with a Normal RNA Replicase," *Science* 153 (1966): 64–67.

8. Brian Goodwin, *How the Leopard Changed Its Spots: The Evolution of Complexity* (New York: Scribner's, 1994), 35–36. So here we have a self-replicating system that evolved into something simpler. One might wonder whether the system also submitted a purchase order for more monomers when the first batch ran out. Now that would be true self-replication!

9. Note that we are not saying that Darwinism requires that evolution proceed toward simplicity. Our point is simply that Darwinism, in itself, does not mandate increasing complexity and inherently favors simplicity. Thus, if we see steadily increasing complexity, something besides selection and variation must be at work. Now, Darwinists have offered rationales for why we might expect increasing complexity strictly on Darwinian grounds (e.g., the irreversibility of certain changes, the complexity cost of arms races, and the lower wall of

complexity below which things are dead). But all such rationales are post hoc—in each case the opposite might well have happened and Darwinism could still be true. Thus, we can imagine (and even program on computer) Darwinian evolutionary scenarios in which reversibility has a selective advantage, in which arms races are won by simplifying, and in which the lower wall of complexity is an absorbing barrier where maximal fitness is conferred by maximal simplicity. Bottom line: Whether evolution in a Darwinian scenario tends toward increasing or decreasing complexity depends on factors other than those dictated by Darwinian theory; moreover, since increased complexity invariably incurs a fitness cost (i.e., there's more to monitor), Darwinian theory inherently favors decreasing complexity.

10. Theodosius Dobzhansky, discussion of G. Schramm's paper in *The Origins of Prebiological Systems and of Their Molecular Matrices*, ed. S. W. Fox (New York: Academic Press, 1965), 310.

Chapter 18: When All Else Fails—Panspermia

1. Fred Hoyle and Chandra Wickramasinghe, *Astronomical Origins of Life—Steps Towards Panspermia* (Dordrecht: Kluwer, 2000).

2. Francis Crick and Leslie Orgel, "Directed Panspermia," *Icarus* 19 (1973): 341–346. Crick also wrote a book on directed panspermia titled *Life Itself* (New York: Simon & Schuster, 1981).

Chapter 19: The Medium and the Message

1. Though note: proof of concept works only when one proves the concept. Origin-of-life researchers are a long way from establishing proof of concept. Indeed, it has completely eluded them. Their willingness to embrace just about any highly speculative scenario for life's origin suggests that in fact they are giving up on proof of concept and acting out of desperation, trying to shore up a materialist explanation of life's origin when life is clearly telling us that its origin is not materialist.

2. Kauffman, *At Home in the Universe*, 31. This passage was written in the mid 1990s. More recently, Kauffman has gone further, denying not just that we don't know the specifics of how life originated but that we don't even have a theory for how life might have originated: "[W]e entirely lack a theory of organization of process, yet the biosphere, from the inception of life to today manifestly propagates organization of process." See Stuart Kauffman, Robert K. Logan, Robert Este, Randy Goebel, David Hobill, and Ilya Shmulevich,

"Propagating Organization: An Enquiry," *Biology and Philosophy* (2007): forthcoming, available online at http://personal.systemsbiology.net /ilya /Publications/BiolPhilosPropagatingOrganization.pdf (last accessed April 23, 2007).

3. Note that this bracket insertion was in the original source for this quote. See the next note.

4. Jason Socrates Bardi, "Life—What We Know, and What We Don't," *TSRI News & Views* (the online weekly of The Scripps Research Institute) 3(30) (October 11, 2004), available online at http://www.scripps.edu/newsandviews/e_20041011/ghadiri.html (last accessed April 23, 2007).

5. George M. Whitesides, "Revolutions in Chemistry" (Priestly Medalist address), *Chemical & Engineering News* 85(13) (March 26, 2007): 12–17, available online at http://pubs.acs.org/cen/coverstory/85/8513cover1.html (last accessed April 23, 2007).

6. *Ibid.*, emphasis added.

7. Davies, *The Fifth Miracle*, 19.

8. Another famous chemist made precisely this point in the mid-20th century: "A book or any other object bearing a pattern that communicates information is essentially irreducible to physics and chemistry. . . . We must refuse to regard the pattern by which the DNA spreads information as part of its chemical properties." Michael Polanyi, "Life Transcending Physics and Chemistry," *Chemical & Engineering News* (August 21, 1967): 62.

9. David Baltimore, "DNA Is a Reality beyond Metaphor," *Caltech and the Human Genome Project* (2000): available online at http://pr.caltech.edu:16080/events/dna /dnabalt2.html (last accessed April 23, 2007).

10. Manfred Eigen, *Steps Towards Life: A Perspective on Evolution*, trans. Paul Woolley (Oxford: Oxford University Press, 1992), 12.

11. Eörs Szathmáry and John Maynard Smith, "The Major Evolutionary Transitions," *Nature* 374 (1995): 227–232.

12. Christian de Duve, *Vital Dust: Life as a Cosmic Imperative* (New York: Basic Books, 1995). This book is divided in seven parts. These are the headings for the first four parts.

13. *Ibid.*, 10. Observe that de Duve's analogy miscarries: the transition from stone to bronze to iron tools and weapons was the work of intelligent designers who knew what problems they wanted to solve. What reason do we have for believing that chemicals have problems they need to solve?

Chapter 20: The God of the Gaps

1. All quotes in this paragraph are from Francis S. Collins, *The Language of God: A Scientist Presents Evidence for Belief* (New York: Free Press, 2006), 92–93.

2. All quotes in this paragraph are from Francis S. Collins, "Faith and the Human Genome," *Perspectives on Science and Christian Faith* 55(3) (2003): 152.

3. All quotes in this paragraph are from Hazen, *Genesis*, 80.

4. See Jonathan Wells, "Darwin of the Gaps: A Review of Francis S. Collins's *The Language of God*," available online at http://www.discovery.org/a/4529 (last accessed May 4, 2008).

Chapter 21: Cellular Engineering

1. Compare molecular biologist James Shapiro's work on "natural genetic engineering." Shapiro is neither a Darwinist nor an ID proponent but regards the ability of cells to do their own engineering as crucial to biological innovation and therefore to evolution. For a representative sample of his work, see http://shapiro.bsd.uchicago.edu /index3.html ?content=genome.html (last accessed April 27, 2007).

2. Hubert Yockey, "Self-Organization Origin of Life Scenarios and Information Theory," *Journal of Theoretical Biology* 91 (1981): 13–16.

3. William A. Dembski, *The Design Revolution: Answering the Toughest Questions about Intelligent Design* (Downers Grove, IL: InterVarsity, 2004), 151–152.

4. For instance, it is a fallacy of composition to argue that because bricks are hard, homogeneous, rectangular solids, therefore houses composed of bricks are likewise hard, homogeneous, rectangular solids.

5. The workings of human body parts must be understood, in some respects, as mechanical in order to be understood at all. For example, when medical doctors advise that you crouch to pick up a heavy load, rather than bend over, they are telling you that your back is not suited to a use as the arm of a crane. People who ignore that fact can damage their backs. True, in that case living tissue gets damaged, but the problem is a mechanical one. The solution is also mechanical: distribute the load to your arms and shoulders, which are suited to that task, and aim it away from your mid and lower back. In the same way, cells are not mere collections of machines; yet some aspects of

their function require an understanding of mechanical principles.

6. For instance, it is a fallacy of division to argue that because houses are dwelling places for humans, therefore the bricks composing houses are likewise dwelling places for humans.

7. See http://www.protolife.net (last accessed April 27, 2007).

8. See http://www.protolife.net/company/profile.php (last accessed April 27, 2007). Emphasis added.

Chapter 22: Irreducible Complexity

1. For a full theoretical account of irreducible complexity, see William A. Dembski and Jonathan Wells, *The Design of Life: Discovering Signs of Intelligence in Biological Systems* (Dallas: Foundation for Thought and Ethics, 2008), ch. 6.

2. See Michael Behe, *Darwin's Black Box: The Biochemical Challenge to Evolution* (New York: Free Press, 1996). In this book, Behe introduced the concept of irreducible complexity and traced it through various systems inside the cell.

Chapter 23: Conservation of Information

1. See Cullen Schaffer, "A Conservation Law for Generalization Performance," *Machine Learning: Proceedings of the Eleventh International Conference*, eds. H. Hirsh and W. W. Cohen (San Francisco: Morgan Kaufmann, 1994), 259–265; Thomas M. English, "Some Information Theoretic Results on Evolutionary Optimization," *Proceedings of the 1999 IEEE Congress on Evolutionary Computation* 1 (1999): 788–795; and William A. Dembski and Robert J. Marks II, "The Search for a Search," 2008, typescript, http://www.EvoInfo.org. The last paper by Dembski and Marks constitutes the most powerful and general formulation of conservation of information at the time of this writing.

2. Dembski and Marks, "The Search for a Search."

3. *Ibid.*

4. Richard Dawkins, *The Blind Watchmaker* (New York: Norton, 1987), 50.

5. *Ibid.*, 13.

6. *Ibid.*, 316.

7. *Ibid.*, 141.

8. Richard Dawkins, *Climbing Mount Improbable* (New York: Norton, 1996).

9. For details, see Dembski and Marks, "The Search for a Search."

Chapter 24: Thinking Outside the Box

1. Paul Davies, *The Fifth Miracle: The Search for the Origin and Meaning of Life* (New York: Simon & Schuster, 1999), 17.

2. See Gerald Kaufman, *The Book of Modern Puzzles* (New York: Dover, 1940), 46.

3. In a fair-minded world, this openness to multiple explanatory options would be welcomed. Materialism, however, is anything but fair-minded, in contrast to intelligent design. ID can honestly admit the possibility that matter has tremendous creative powers sufficient to bring about life—it's just that when we look at the world, matter displays nothing of the sort. But materialism cannot admit the possibility that intelligence preceded and was responsible for life. If it could be shown beyond reasonable doubt that without intelligence matter could not organize itself to form life, the materialist would declare the problem of life's origin insoluble. For the materialist, a solution to life's origin that requires intelligence is no solution at all.

Chapter 25: A Reasonable Hypothesis

1. Charles S. Peirce, "How to Make Our Ideas Clear," *Popular Science Monthly* 12 (January 1878): 286–302.

2. Michael Ruse, *Can a Darwinian Be a Christian?* (Cambridge: Cambridge University Press, 2001), 64. Emphasis added.

3. To watch a webcast of the entire debate, go to http://www.bu.edu/com/greatdebate /fall05/index.html (last accessed April 30, 2007).

4. "It is not implausible that life emerged as a phase transition to collective autocatalysis once a chemical minestrone, held in a localized region able to sustain adequately high concentrations, became thick enough with molecular activity." Kauffman, *At Home in the Universe*, 274.

5. See Richard Dawkins, *The God Delusion* (New York: Houghton Mifflin, 2006).

6. Richard Dawkins, *River Out of Eden: A Darwinian View of Life* (New York: Basic Books, 1995), 98.

Epilogue: Atheism As A Speculative Faith

1. Indeed, what is atheism's record of accomplishment? Dinesh D'Souza, responds to Christopher Hitchens's claim that "religion poisons everything" by noting: "Religion didn't poison Dante or Milton or Donne or Michelangelo or Raphael or Titian or Bach! Re-

ligion didn't poison those unnamed architectural geniuses who built the great Gothic cathedrals. Religion didn't poison the American founders who were for the most part not Deist but Christian. Religion didn't poison the anti-slavery campaigns of William Lloyd Garrison or William Wilberforce, or the civil rights activism of the Reverend Martin Luther King. The real question to ask is, what does atheism offer humanity? In Tonga, as in America, the answer appears to be: Nothing." See Dinesh D'Souza, "What Has Atheism Done for Us?" *Townhall* (October 31, 2007): available online at http://www.town-hall.com/columnists/DineshDSouza/2007/10/31/what_has_athe-ism_done_for_us (last accessed May 5, 2008).

2. See David Berlinski, *The Devil's Delusion: Atheism and Its Scientific Pretensions* (New York: Crown Forum, 2008).

About the Authors

William A. Dembski is a Senior Fellow with the Discovery Institute's Center for Science and Culture. He has authored or edited more than a dozen books, including the first book on intelligent design to be published by a major university press, *The Design Inference: Eliminating Chance through Small Probabilities* (Cambridge University Press, 1998). He has seven earned degrees, including two doctorates, one in philosophy from the University of Illinois at Chicago, the other in mathematics from University of Chicago. His work has been featured on the front page of the *New York Times* and he has appeared on numerous radio and television broadcasts, including ABC's *Nightline* and Jon Stewart's *The Daily Show*. Bill and his wife, Jana, live in Central Texas with their daughter and twin boys.

Jonathan Wells is a Senior Fellow with the Discovery Institute's Center for Science and Culture. He holds two doctorates, one in molecular and cell biology from the University of California at Berkeley, the other in religious studies from Yale University. As the author of *Icons of Evolution: Why Much of What We Teach about Evolution Is Wrong* (Regnery, 2000), Wells has emerged as one of the key figures for reforming the teaching of evolution by correcting textbook errors and by insisting that the evidence that both confirms and disconfirms Darwinism be taught. He is a widely acclaimed public speaker who, through his lectures and debates, inspires a younger generation of scholars to develop intelligent design as a fruitful scientific research program. Jonathan and his wife, Lucy, live in the Seattle area.

In a controversial satirical documentary released in April 2008, author, former residential speechwriter, economist, lawyer, and actor Ben Stein travels the world, looking to some of the best scientific minds of our generation for the answer to the biggest question facing all Americans today:

"Are we still free to disagree about the meaning of life? Or has the whole issue already been decided . . . while most of us weren't looking?"

The debate over evolution is confusing and to some, bewildering: "Wasn't this all settled years ago?" The answer to that question is equally troubling: "Yes . . . and no."

Expelled: No Intelligence Allowed presents a point-of-view so powerful, that it literally forces a re-examination of these issues

Go to www.expelledthemovie for access to educational resources and for information on how to purchase your own copy of *Expelled: No Intelligence Allowed* on DVD.